MAP PROJECTIONS

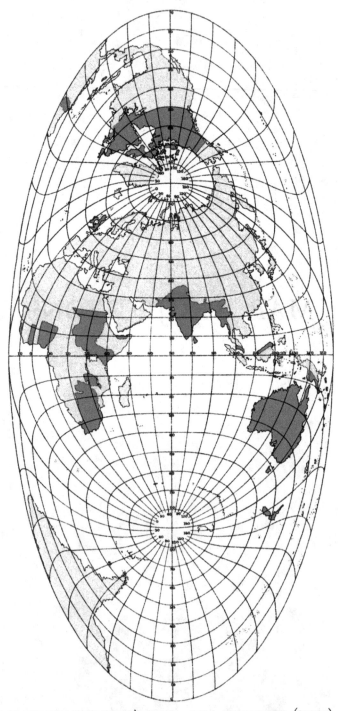

TRANSVERSE MOLLWEIDE'S EQUAL-AREA PROJECTION (CLOSE)

MAP PROJECTIONS

BY

ARTHUR R. HINKS, C.B.E., M.A., F.R.S.

SECRETARY OF THE ROYAL GEOGRAPHICAL SOCIETY

SECOND EDITION. REVISED AND ENLARGED

CAMBRIDGE

AT THE UNIVERSITY PRESS

1921

CAMBRIDGE
UNIVERSITY PRESS

University Printing House, Cambridge CB2 8BS, United Kingdom

Cambridge University Press is part of the University of Cambridge.

It furthers the University's mission by disseminating knowledge in the pursuit of education, learning and research at the highest international levels of excellence.

www.cambridge.org
Information on this title: www.cambridge.org/9781316611791

First edition 1912
Second edition 1921
First paperback edition 2016

A catalogue record for this publication is available from the British Library

ISBN 978-1-316-61179-1 Paperback

PREFACE TO FIRST EDITION

THE subject of Map Projections has become over complicated because it has interested many mathematicians. The parts of the subject which are of mathematical interest are embodied in theorems of great elegance and considerable difficulty; in particular, the theory of the conformal representation of one surface upon another leads to developments which, from the mathematical standpoint, are of the highest importance, and possess an extensive literature. Yet we shall find that the property of orthomorphism, which plays such a large and difficult part in the theory of Map Projections, is not in most cases of any great advantage or importance in actual mapmaking. [A*]

In writing a book on Map Projections, the usual course has been to present the general mathematical theory first, and to discuss the practical questions involved at a later stage. The result is that the geographer sometimes finds himself unable to follow the bearing of the mathematics, and arrives at the consideration of the practical side of the subject in a very unformed state of mind.

I propose to adopt the principle of a very distinguished topographer, that in a book on Map Projections intended for the mapmaker and the map user, "one should draw the line at the root of minus one." If one follows this course, one often finds it impossible to show how some of the more elegant projections may be arrived at by a deduction from general principles. When, however, the formula of the projection is once given, it is always easy to work backwards and to demonstrate its properties. The amount of mathematics required in the two cases is very different.

It will be seen, therefore, that this book has no pretensions to consideration as a treatment of the theory of projections from the mathematical point of view. On the contrary, its object may be stated very briefly as follows: There are some thirty map projections of importance, of which about half are in more or less general use. All of them have certain valuable properties, and equally serious defects. It is important to have a clear graphical or numerical idea of the merits and defects of each; to be able to decide at once on its suitability for a given map; or when one finds it actually employed on a map, to recognise what a map so constructed will do, and what it will not do.

I shall try in this book to make clear the relations between the various projections; the extent to which they possess the qualifications which a good map projection should possess; the methods by which they can be constructed; and the way in which maps so constructed can be used. The last matter is of considerable present importance. Relatively few people have to make maps, but very many have to use them, and it is necessary to learn to guard against fatal mistakes. The introduction of Map Projections into the schedule of Geography for the examination for First Class appointments in the Home and Indian Civil Service is a welcome recognition of this fact. In preparing this slight account of a large and diffuse subject I have had the advantage of many discussions with Colonel Close, C.M.G., R.E., Director-General of the Ordnance Survey, to whom I am indebted for my first acquaintance with its beauties. In making the calculations of the numerical properties of the various projections in use I have had the help of several pupils in the Cambridge Geography School; and in particular I have to thank Mr F. M. Deighton, B.A., of Trinity College, and Mr T. W. Glare, B.A., of Sidney Sussex College, Cambridge, for the great assistance they have given me.

I have not tried to give a set of drawings of all the projections treated in this book. Only very small scale diagrams would have been possible, and these cannot do justice to the projections that are most suitable for topographical and for Atlas maps. In their useful central regions these are very similar to one another until they are measured up, and the eye can hardly distinguish between them; when they are constructed as world-maps, for which most of them are entirely unsuitable, they become easily distinguishable and at the same time absurd. Thus, to represent the whole sphere upon a conical or polyconic projection is to obscure the real merits and the proper uses of the projection. I have therefore been content to give references to the places where these projections may be seen in actual use and studied on an adequate scale; and have confined myself here to plates of the two or three projections that make good small scale diagrams of the sphere; and to a few explanatory figures. The transverse Mollweide's projection is a new and interesting world-map; I have to thank its inventor, Colonel Close, for permission to include it as the frontispiece of this book. The plate was printed in the Geographical Section of the General Staff, by kind permission of Colonel Hedley, R.E.

A. R. H.

CAMBRIDGE,
August 1912.

PREFACE TO SECOND EDITION

WHEN this book was written in 1912, for the use of my pupils in Geography in the University of Cambridge, my knowledge of Map Projections was academic. During the intervening years I have learned much by experience at the Royal Geographical Society, especially during the years of war; and I cannot in 1920 endorse quite all the confident judgments of 1912. For instance, the requirements of the artillery in modern war have brought into great prominence the advantages of an orthomorphic projection for the large scale tactical maps used in stationary warfare; and what I said of orthomorphism in 1912 needs modification.

A few mistakes have been corrected; but otherwise the body of the book stands as it was in the first edition, save for references to notes in modification of judgment, or in amplification of treatment. These notes (indicated in the text thus: A*, B* and so forth) are collected in a new Chapter XI. Other new Chapters deal with the projections for topographical maps on a large scale, which were quite inadequately treated in the first edition; with the calculation of rectangular coordinates and grids for tactical maps; with the history of Map Projections; and with some additional tables.

I am especially indebted to Mr A. E. Young, A.M.I.C.E., formerly Deputy Surveyor General of the Federated Malay States, for much help during the last four years. His investigation of various points which arose during the war led to his writing a work on Map Projections containing original and valuable ideas, which has been recently published by the Royal Geographical Society. I owe to him much of the information given in my additional chapters. The frontispiece, due to Colonel Sir Charles Close, K.B.E., F.R.S., now Director General of the Ordnance Survey, has been reprinted from the War Office stones by kind permission of Colonel E. M. Jack, C.M.G., D.S.O., Chief of the Geographical Section, General Staff.

A. R. H.

LONDON,
January 1921.

CONTENTS

CHAPTER I

INTRODUCTION

CHAPTER II

THE PRINCIPAL SYSTEMS OF PROJECTIONS

CHAPTER III

CYLINDRICAL PROJECTIONS

CHAPTER IV

ZENITHAL PROJECTIONS

CHAPTER V

ZENITHAL PROJECTIONS, *continued*

CHAPTER VI

MODIFIED CONICAL AND CONVENTIONAL PROJECTIONS

CHAPTER VII

PROJECTIONS IN ACTUAL USE

CHAPTER VIII

THE SIMPLE MATHEMATICS OF PROJECTIONS

CHAPTER IX

THE NUMERICAL ERRORS OF PROJECTIONS

CHAPTER X
TABLES

APPENDIX TO FIRST EDITION

CHAPTER XI
CORRECTIONS AND ADDITIONS TO PREVIOUS CHAPTERS

CHAPTER XII
PROJECTIONS FOR TOPOGRAPHICAL MAPS ON LARGE SCALES

CHAPTER XIII
NOTES ON THE HISTORY OF MAP PROJECTIONS

CHAPTER XIV
ADDITIONAL TABLES

CHAPTER I

INTRODUCTION

WE have upon the nearly spherical surface of our globe an arrangement of features whose relative positions, sizes, and shapes we desire to represent as well as possible upon a flat sheet, the map. A perfect representation is impossible, since a plane surface cannot be fitted to a spherical. But there are many different ways of obtaining an approximate representation, whose theory and properties constitute the subject of Map Projections.

Definition of a projection.

The positions of points upon the Earth are for convenience defined by reference to the meridians of longitude and parallels of latitude. Hence if we can find a way of representing the parallels and meridians upon our sheet, we can lay down the points in their positions relative to these lines, and make our map. Any such representation of meridians and parallels upon a plane is a map projection.

It is evident, from our definition, that we use the word Projection in a sense much wider than that which geometry gives it. The majority of map projections are not projections at all in the geometrical sense, and various attempts have been made to find a better word to describe the network of meridians and parallels. But no one has been successful. Map "construction" implies rather too much. The excellent word "graticule" has scarcely established itself, though it is perhaps less open to objection than any other. We shall, then, continue to use the word projection, with the warning that it is not to be interpreted as meaning a geometrical projection. Strictly geometrical projections are of very little use in map making, and it is a mistake to begin by considering the few that are used, and to proceed afterwards to the very many useful projections which are not derived from the sphere by any perspective construction.

The number of possible ways of constructing a projection is infinite, even if we restrict our definition to the statement that any *orderly* construction of meridians and parallels may be considered a projection. It is obvious, however, that all these constructions are not equally good; and to test the merits and defects of a projection we must consider what properties it should possess in order to be useful.

We shall find that map projections are to be judged by the following criteria:

(1) the accuracy with which they represent the scale along the meridians and parallels.

(2) the accuracy with which they represent areas.

(3) the accuracy with which they represent the shape of the features of the map.

(4) the ease with which they can be constructed. [B*]

We will consider these criteria in order.

The representation of scale.

The scale of a map in a given direction at any point is the ratio which a short distance measured on the map bears to the corresponding distance upon the surface of the Earth.

We must limit our definition to *short* distances because the scale of a map will generally vary from point to point; hence in defining scale we must confine ourselves to small elements of distance in the way which is familiar to every beginner in the differential calculus.

We must also be careful to see that we are comparing distances in directions which really correspond, the one to the other, upon the Earth and upon the map. The meridians and parallels all over the Earth cut one another at right angles. But there are many map projections in which they do not cut one another at right angles, and in consequence two directions at right angles upon the Earth do not necessarily correspond to two directions at right angles upon the map. We shall avoid confusion if we confine ourselves as much as possible to the consideration of scale along the meridians and the parallels of the map, which necessarily correspond to the meridians and parallels of the Earth.

It would of course be desirable that the scale of the map should be correct in every direction at every point. If it were, the plane map would be a perfect representation of the spherical surface, and could therefore be fitted to it. But this is impossible. Hence the scale of a map cannot be correct all over the map.

We can, however, choose a projection in which the scale in a certain direction, say along the meridian, or along the parallel, is correct at every point of the map. But in this case the scale in any other direction will be wrong at most points. And one of our objects will be to keep this necessary error as small as possible.

The representation of areas.

For some purposes, especially political and statistical, it is important that areas should be represented in their correct proportions. A projection which does this is called an equal area projection, or an equivalent projection. We shall use the former name in this book.

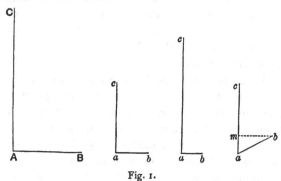

Fig. 1.

Suppose that AB, AC are two short distances at right angles to one another at any point on the Earth. If the corresponding distances ab, ac upon the map were always in the same proportion and also at right angles to one another, the projection would clearly be an equal area projection. But these conditions cannot be fulfilled, for if they were fulfilled at every point, the map would be a perfect map, which is impossible.

There are however two distinct ways in which the equality of areas may be preserved upon the map.

(α) *ab, ac* may still be at right angles, but with the scale of one increased and of the other decreased, in inverse proportion.

(β) or *ab, ac* may be no longer at right angles; but while the scale of *ac* is maintained correct, that of *ab* is increased in such a proportion that the perpendicular distance of *b* from *ac* is correct.

It is clear that in either case the projection is equal area.

The representation of shape.

The representation of shape as nearly correctly as possible is perhaps the most important function of a map. It is evidently not possible to represent the shape of a large country correctly upon a map, for if it were, the map would be perfect, which is impossible.

But if at any point the scale along the meridian and the parallel is the same (not correct, but the same in the two directions) and the parallels and meridians of the map are at right angles to one another, then the shape of any very small area on the map is the same as the shape of the corresponding small area upon the Earth. The projection is then called *orthomorphic* (right shape).

But it is important to notice the restriction to *very small areas*. Since the scale necessarily varies from point to point, big areas are not correctly represented. Hence it is clear that the term orthomorphic must be used in a carefully limited sense. It has, in fact, a mathematical significance and interest which is apt to be of little use in practice. [A*]

For example, suppose we had a strip of country a mile wide, along a meridian, and divided into two equal parts by parallels of latitude (Fig. 2 *a*). A projection which is orthomorphic in the mathematical sense might represent the figure thus (Fig. 2 *b*). It will be noticed that all the angles are preserved in their true magnitudes as right angles, but the strip on the map is no longer of uniform width, it is no longer divided into equal parts, and it is unsymmetrical.

Another orthomorphic projection might represent the same strip thus (Fig. 2 *c*): the angles are preserved as before, the areas are modified so that the strip is no longer bisected; but there is symmetry and no general bending.

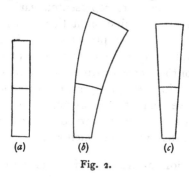

(*a*)   (*b*)   (*c*)

Fig. 2.

It is clear that the latter projection may be much superior to the former in representing the shapes of considerable areas, especially for countries having a great extent in latitude. The essential difference is, that in the former case the projections of the meridians are not straight lines, while in the latter case they are.

We shall find that an orthomorphic projection is generally not much use for map making unless the meridians are straight lines on the map. [C*]

We shall also notice that projections which are not, mathematically speaking, orthomorphic may often represent the shape of large areas better than is done by projections that are orthomorphic.

We shall therefore be prepared to find that the mathematical property of orthomorphism does not always, or indeed usually, give the map any considerable practical advantages; and that orthomorphism may generally be sacrificed to other less theoretically elegant, but more useful properties.

We may define orthomorphic projections by either of two properties:

(α) At any point the scale in all directions is the same. (This is the definition used above.)

Or (β) The angles in any small figure on the Earth are preserved unaltered on the map. (It is geometrically obvious that this definition is equivalent to the preceding.)

It follows from either of these definitions that orthomorphic projections preserve the shapes of small areas unaltered, though the scale on which they are represented varies from point to point upon the map. At first sight this preservation of shape appears to be an important property. It should however be remembered that, so long as we do not attempt to represent too large a fraction of the whole Earth upon one map, a great many of the usual projections are pretty nearly orthomorphic for small areas; while if we remove the restriction to small areas the general shape is often better preserved in projections which are not orthomorphic than in those which are.

The representation of true bearings and distances.

We have just seen that the definition of orthomorphism restricts the property to very small areas; and that the existence of perfect orthomorphism, according to definition, is no guarantee even for the approximate preservation of the shape of large configurations.

We want some criterion for the degree of success of a given projection in preserving the shape of a country from gross distortion, and we shall find it useful to consider how far the true bearings and distances from point to point are preserved. For example, we have a map of Europe on a given projection, and we enquire: What is the percentage error of the representation of the distance from Hanover to St Petersburg; or what is the error in the azimuth of this line. Such questions cannot be answered by considering very small areas.

There is a class of projections sometimes named *azimuthal*, from the fact that the azimuths, or true bearings, *from the centre of the map*, of all points, are shown correctly. One of these azimuthal projections also shows distances from the centre of the map correct, and is called the azimuthal equidistant projection. We shall find it useful to ask of each projection, how nearly does it approximate to the azimuthal equidistant, and further, how well does it preserve azimuths and distances, not only from the centre, but from any other point?

The objection to the term *azimuthal* is that it is hard to pronounce, and several writers have followed Germain in calling always this class of projection *zenithal*. Since their most prominent and valuable property is the preservation unaltered of azimuths, or true bearings, from the centre, it appears to the writer that the former name is preferable to the latter, and that it is unfortunate that *zenithal*, which has no very clear meaning, should replace *azimuthal*, whose meaning is precise. We shall not, however, try to return in this book to the older fashion.

Ease of drawing.

This is a property which is theoretically uninteresting, but which is, in practice, of extreme importance. As a general rule, projections which are not built up of straight lines and circles are hard to draw. This rule excludes at once all the strictly geometrical projections, except the stereographic, which is built entirely of circles. [D*]

Further, arcs of circles of very large radius are hard to draw; and for this reason graphical constructions often break down somewhere or other, requiring circles too large for the drawing table.

In such a case a series of points on the circle must be computed or constructed graphically. Hence the formulae of computation become very important.

We shall ask ourselves at the end of the section on each projection: Is it easy to draw, or can tables for it be constructed easily?

The choice of projection for a map.

It is clear that we cannot say anything in detail upon this subject until we have examined the properties of the principal projections in use. But we shall do well to bear in mind from the beginning that there are three broad classes of maps:

(a) Maps of the whole world or of a hemisphere, on one

sheet. We may call any map that represents a hemisphere, or more, a World Map. These will always be on small scales.

(β) Maps which show a considerable portion of the Earth, such as a continent, but not a whole hemisphere. These will also be on a small scale. We shall find it convenient to call them Atlas Maps.

(γ) Maps on a comparatively large scale, each representing in detail a fairly small area of country. We may call these Survey Maps.

The projections for maps in sections (α) and (β) will generally be constructed independently for each map, and there will be no question of adjacent maps fitting. But it may be thought desirable that the sheets of a Survey map should fit together, so that they may be combined to form larger maps if necessary. This will require that the projection for the whole survey should be determined, and that each sheet should not be plotted independently but should be a definite part of this projection. We shall see that this is practicable only for a small country like Great Britain, and has in any case considerable disadvantages.

It is evident, however, that in the *use* of Survey maps, the question of projections does not often arise. Each sheet covers so small a portion of the whole surface of the Earth that it is practically a perfect representation, if the original choice of a projection for the Survey has been well made, and more particularly, if the perpendicularity of meridians and parallels is preserved. This reservation is of prime importance. [E*]

Atlas maps cannot be treated as practically errorless. In these the errors in scale, area, and shape become considerable, and are unavoidable. We cannot make measurements upon the map until we know the errors which are due to the projection. We shall therefore arrange our detailed consideration of projections so as to give the common Atlas projections as much prominence and priority as is consistent with an orderly development of the subject.

CHAPTER II

THE PRINCIPAL SYSTEMS OF PROJECTIONS:
CONICAL PROJECTIONS

THE greater number of useful projections for Atlas maps belong to one or the other of two great classes, the Conical (including the cylindrical) and the Zenithal projections. These names describe the method of construction. It is useful to give each projection a second name which describes its principal property, such as equal area, orthomorphic, and so on. We shall therefore describe projections as the Conical equal area, the Zenithal orthomorphic, the first, or generic name, describing its construction, the second, or specific name, its most important property. When the name of the inventor, Lambert or Gauss, Sanson or Delisle, is usually associated with the projection, we may give it in brackets, thus: Conical orthomorphic with two standard parallels (Lambert's second, or Gauss').

We shall find, however, that this principle of nomenclature cannot be made to cover all cases without some appearance of pedantry, and that there are well known projections, such as Mercator's or the Stereographic, which will be treated in their systematic places but referred to generally by their simple names. Thus the Stereographic projection is a zenithal orthomorphic; but as it is one of several which can be thus named, it is convenient to call it simply the Stereographic.

We shall find, also, that this way of naming projections is convenient, rather than consistently logical. For zenithal projections are, from one point of view, only special cases of conical. Moreover, all zenithal projections are azimuthal, so that one of the principal properties of this class of projection is implied in its generic name. Thus a zenithal equal area projection has two important properties: it is both azimuthal and equal area. But as we become familiar with the subject this want of strict consistency in the nomenclature will not be a serious difficulty.

We shall find it best to defer the consideration of the classification of map projections until we have become acquainted with the more obvious properties of each projection. It will then be evident that a logical classification requires in the first place a re-consideration of the names which are given to the projections. Inasmuch however as a disturbance of the accepted names, even though they are unsystematic, would certainly create more confusion than it would remove, we shall find it best to retain the generally accepted names, taking Germain's *Traité des projections* as our standard authority, and shall make suggestions for a more logical nomenclature only in our discussion of classification, without attempting to introduce any reform of such doubtful advantage into the body of the work.

Conical projections.

In all the usual conical projections the meridians are straight lines converging to a point, the vertex, and the parallels are concentric circles described about that point.

The meridians are equally spaced, and make with one another angles which are a certain fraction n of the angles which the corresponding terrestrial meridians make with one another at the poles. The quantity n we will call the constant of the cone. It must lie between the values 0 and 1.

The spacing of the parallels depends upon the particular property which we wish the projection to fulfil.

One parallel, and sometimes a second, is made of the true length; that is to say, if the map is to be on the scale of one-millionth, the length of the complete parallel on the map will be one-millionth of the corresponding terrestrial parallel. This is called a Standard parallel.

Simple conical projection with one standard parallel.

Suppose that we begin by constructing the Conical projection with one standard parallel. If R is the radius of the Earth (supposed spherical) and ϕ the latitude of the selected parallel, the length of the parallel is $2\pi R \cos \phi$.

Describe with radius $R \cot \phi$ an arc whose length is $2\pi R \cos \phi$, the length of the standard parallel. This arc will subtend at the centre of the circle an angle $2\pi \sin \phi$, and the constant of the cone is $\sin \phi$.

Divide the arc into 36 parts, and join each dividing point to the centre. These lines will represent the meridians at intervals of 10°.

Along any one of these meridians lay off distances above and below the standard parallel equal to the distances from the

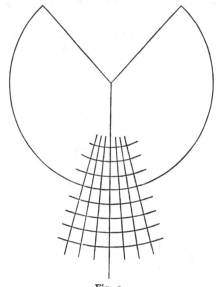

Fig. 3.

standard parallel to each parallel of even ten degrees, the distances being reduced to the scale of the map. Describe circles through the points so obtained, concentric with the standard parallel. These will be the parallels of our projection at intervals of 10°.

Consider briefly the properties of this projection.

The scale along every meridian is correct, as is the scale along the standard parallel. It is easy to show that the scale along any other parallel is too great, and that the error increases as we get further away from the standard parallel. [F*]

The projection is evidently not equal area; for the meridians and parallels are at right angles to one another, and the scale is correct along the meridians and wrong along all parallels but one. The percentage error in the representation of any small area is evidently the same as the percentage error of the scale of the parallel through it.

The projection is evidently not orthomorphic, since the scale is not the same in all directions at any point. The projection is easy to draw, unless difficulty is found in describing circular arcs of very large radius, and straight lines converging to a distant point.

The projection is evidently good for any extent of longitude along the standard parallel. But north and south of it the

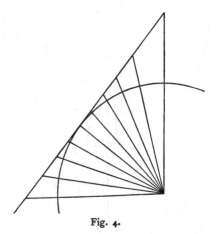

Fig. 4.

errors of scale along the parallel continually increase; and the parallel of latitude 90°, the pole of the hemisphere containing the standard parallel, is represented not by a point but by a circle. This shows that the projection is unsuitable for representing high latitudes; the smaller extent in latitude it has, the better.

It is easy to see geometrically that if the complete projection were cut out and rolled into a cone, it could be placed to touch a model of the Earth, constructed on the scale of the map, along the standard parallel.

On this account it is usual in text books to begin the study of this projection by supposing the tangent cone constructed, and afterwards cut along a generator and laid out flat. But this way of regarding it is a little dangerous, since it may lead the beginner to suppose that the projection is made by geometrically projecting the parallels and meridians of the Earth on to a tangent cone.

The meridians may indeed be obtained in this way by projection from the centre of the Earth; but the parallels cannot be, for they encircle the cone at their true distances apart; and it is evident that one cannot project a series of parallel circles on the sphere into a system of parallel circles the same distance apart upon the tangent cone. This is easily seen by a study of Fig. 4.

The simple conical projection is very much used in Atlas maps of countries not too large.

Its principal defect is the increasing error of scale along the parallel as one leaves the standard parallel. This can be much diminished by making two standard parallels instead of one.

The conical projection with two standard parallels and true meridians.

Suppose that we select two parallels, one towards the top and one towards the bottom of our map; and make them both the correct length, and the true distance apart. Given these conditions we can easily find the radii of the circular arcs for the two standard parallels (see Chap. VIII). And when these are found, the meridians and the other parallels may be constructed as in the preceding case.

The scale is correct along all the meridians, and along the selected standard parallels. It is easy to show that between the standard parallels the scale along the parallel is too small; outside it is too large, as in the preceding projection. But the error does not increase nearly so rapidly as it does in the former case. Hence the conical projection with two standard parallels is much better than that with only one standard parallel, for all maps with a considerable extent of latitude. [G*]

The same considerations as before show that the projection

is neither equal area nor orthomorphic.   But it is more nearly so than the preceding, since the errors of scale are smaller.

It is easy to draw.

As before, the pole is represented by a parallel of finite length, which makes it clear that the projection is useless in extreme latitudes.

This is one of the best of all projections for an Atlas map, but its merits have been very much overlooked until recently, when it was adopted for the original one in a million maps of Great Britain and of India, only to be displaced by the resolutions of the International Map Committee (see page 56).

This projection is commonly, but quite wrongly, called the "secant conic" projection, as distinguished from the simple or "tangent conic" projection described above.   But it is clear that the description is false.   If we passed a cone through the two selected standard parallels upon the sphere, cut it along a meridian, and laid it out flat, it would not give the same projection as that which we are now considering.   The two standard parallels would be of the right length, but they would evidently not be at the right distance apart; in fact the distance between them would be a *chord* of the sphere instead of the corresponding arc.   The beginner must take great care to avoid this mistaken idea of the nature of the conical projection with two standard parallels.

The conical equal area projection with one standard parallel.

In all conical projections the scale along the parallels is incorrect, except for the standard parallel or parallels.   Hence if we wish to make an equal area conical projection we must suitably modify the distances between the parallels, and make the scale along the meridians incorrect in the inverse proportion.

Suppose, for example, that we wish to modify the simple conic with one standard parallel, to make it equal area.

Let ϕ_0 be the latitude of this parallel.   We shall show later (Chap. VIII) that we may attain the desired result by the following means:

Keep the standard parallel its correct length, but describe it with radius

$2R \tan \frac{1}{2} (90° - \phi_0)$ instead of $R \cot \phi_0$ or $R \tan (90° - \phi_0)$.

The cone is no longer the tangent cone at the standard parallel. And the constant of the cone becomes $\cos^2 \frac{1}{2} (90° - \phi_0)$ instead of $\sin \phi_0$.

Instead of marking off the other parallels at their true distances, describe them with radii computed by the formula

radius for lat. $\phi = 2R \sec \frac{1}{2} (90° - \phi_0) \sin \frac{1}{2} (90° - \phi)$.

We shall prove later (*loc. cit.*) that this construction gives an equal area projection.

The scale along both meridians and parallels is wrong, in inverse proportions. It is consequently less orthomorphic than either of the two preceding projections, in which the scale along the meridians was correct.

It is easy to compute and as easy to draw as other conical projections.

Unlike the preceding conical projections, the pole is not represented by a parallel, but by a point, the vertex. This does not, however, make it any more suitable for the representation of the polar regions, the whole of which are included on the map in a sector of angle $2n\pi = 2\pi \cos^2 \frac{1}{2} (90° - \phi_0)$.

On the side of the standard parallel towards the vertex the scale along the meridians increases quickly; and since the projection is equal area, it follows that the scale along the parallels decreases in inverse proportion. On the opposite side of the parallel the reverse takes place. In the former case the configurations are elongated north and south; in the latter case east and west. Hence the projection is not useful for a map having great extension in latitude; and it has been employed very little.

We should note that, though the above is what is always known as *the* conical equal area projection, it is not by any means the only one possible. Without altering the constant of the cone, or the angles between the meridians, one may so modify the distances between the parallels that the scale along the meridians is inversely proportional to the scale along the parallels. The projection is then equal area; and since a cone of any angle may be treated in this way, there is an infinite class of conical equal area projections, of no particular value or interest.

The conical equal area projection with two standard parallels (Albers').

By a different modification we can obtain a conical equal area projection with two standard parallels, as follows.

Let ϕ_1, ϕ_2 be the latitudes of the parallels chosen as standard, and R the radius of the Earth.

The standard parallels are described of their correct lengths, but with radii r_1, r_2 given by

$$r_1 = kR \cos \phi_1, \quad r_2 = kR \cos \phi_2,$$

where
$$k = \operatorname{cosec} \tfrac{1}{2}(\phi_1 + \phi_2) \sec \tfrac{1}{2}(\phi_1 - \phi_2).$$

And the radius r of any other parallel ϕ is given by

$$r^2 = 2kR^2 (\sin \phi_1 - \sin \phi) + r_1^2,$$

or
$$= 2kR^2 (\sin \phi_2 - \sin \phi) + r_2^2.$$

The constant n of the cone is

$$1/k = \sin \tfrac{1}{2}(\phi_1 + \phi_2) \cos \tfrac{1}{2}(\phi_1 - \phi_2),$$

or
$$\tfrac{1}{2}(\sin \phi_1 + \sin \phi_2).$$

We shall show in Chap. VIII that this gives an equal area projection.

It is rather troublesome to compute, but as easy to draw as other conical projections.

The scale along both parallels and meridians is incorrect, and it is not so orthomorphic as the ordinary conic with two standard parallels.

Outside the standard parallels the scale along the meridians is too small, and the defect becomes more pronounced the further the extension of the map north and south. The pole is represented by a parallel. Within the standard parallels the scale along the meridians is too large. Since the projection is equal area, the scale along the parallels is inversely proportional to the scale along the meridians.

This projection has been used for maps of Russia, and also for the Austrian General Staff map of Central Europe.

Conical orthomorphic projection.

All conical projections have their meridians and parallels intersecting at right angles, which is the first essential for orthomorphism. But in the projections already described the scale along the meridians and parallels is different. If they can be modified so as to make the scale the same in these two perpendicular directions (and consequently in all directions) at any point, they will become orthomorphic.

Let χ be the co-latitude ($=90°-$ the latitude) of a parallel; and r its radius upon the projection. It is easy to show that if r is computed by the formula

$$r = m \, (\tan \tfrac{1}{2}\chi)^n,$$

the projection is orthomorphic. [H*]

m is a constant which defines the scale. We will consider its value later.

n may have any value between 0 and 1; and by varying n we have a whole series of conical projections, all of which are orthomorphic. The angle between two meridians upon the projection is n times the angle between them on the Earth; and we call n the constant of the cone.

We can choose the constant m so that any desired parallel is its true length.

If the parallel of co-latitude χ_0 is to be the standard parallel, it is easy to show that

$$m = \frac{R \sin \chi_0}{n \, (\tan \tfrac{1}{2}\chi_0)^n}.$$

We still have the constant of the cone, n, at our disposal.

If we choose it so that the cone has the same angle as the cone of the simple conic projection, which will touch the Earth along the standard parallel, then

$$n = \cos \chi_0.$$

The conical orthomorphic projection with one standard parallel may therefore be constructed from the formula

$$r = R \frac{\tan \chi_0}{(\tan \frac{1}{2}\chi_0)^{\cos \chi_0}} \cdot (\tan \frac{1}{2}\chi)^{\cos \chi_0}.$$

The scale will be correct in all directions at points along the standard parallel (for the projection is orthomorphic). Elsewhere the scale will be too large.

As the scale increases on each side of the standard parallel, it is clear that there will be pairs of parallels, one on each side of the standard, for which the scale is the same. And it is easy to show that one such pair of co-latitudes χ_1 and χ_2 are connected by the relation

$$n = \frac{\log \sin \chi_1 - \log \sin \chi_2}{\log \tan \frac{1}{2}\chi_1 - \log \tan \frac{1}{2}\chi_2}.$$

The scale at points along this pair of parallels is the same, but it is not correct, since it differs from the scale along the chosen standard parallel. But we have only to change the scale constant of the map, and we have evidently a conical orthomorphic projection with *two* standard parallels instead of one.

The two maps are precisely similar except in scale. All questions of deformation and variation of scale are the same upon the two. But it is an obvious advantage to the general accuracy of scale of the map to have two parallels standard instead of one. Hence when the conical orthomorphic projection is used, it is always that with two standard parallels, which is Lambert's second, or Gauss'.

The conical orthomorphic projection with two standard parallels.

The two parallels which are to be standard are chosen, of co-latitudes χ_1 and χ_2. The constant of the cone, n, is given by the relation

$$n = \frac{\log \sin \chi_1 - \log \sin \chi_2}{\log \tan \frac{1}{2}\chi_1 - \log \tan \frac{1}{2}\chi_2},$$

and the radii, as before, are given by

$$r = m (\tan \frac{1}{2}\chi)^n,$$

where $m = R \sin \chi_1/n (\tan \frac{1}{2}\chi_1)^n$ or $R \sin \chi_2/n (\tan \frac{1}{2}\chi_2)^n$.

It is easily seen that *m* is the value of *r* for the equator.

In these conical orthomorphic projections the pole is a point—the vertex of the cone.

The scale increases north and south. If one parallel is standard, the scale at any point not on this parallel is too great. If two parallels are standard, the scale between them is too small, and outside them is increasingly too great. Hence like all other conical projections, they are unsuitable for maps having a great extension in latitude. For maps in which the range of latitude is not too great the scale error can be kept fairly small when two selected parallels are standard. Hence areas are well represented. And in spite of the apparent complication of the expressions for *n* and *r* the projection with two standard parallels is easy to compute, and no harder to draw than any other conical projection. It makes an excellent map of a country like Russia, but has not come into general use, except in Debes' Atlas.

General remarks on the conical projections.

We have seen that the true conical projections have a range of properties sufficiently wide to make them appear, at first sight, a very useful and valuable family—they may be made nearly true to scale over a fairly wide area, and consequently nearly equal area and nearly orthomorphic; or they may be made precisely equal area and less orthomorphic; or precisely orthomorphic and less equal area. But the fact remains that the equal area and orthomorphic projections have been used very little; while until lately the ordinary conic with two standard parallels has been almost equally neglected. There remains the true simple conic, which has been very much used for small atlas maps.

In the following chapters we shall have to investigate in greater detail the numerical properties of these different projections, and we shall find at any rate a partial explanation of these facts. For small countries it will appear that the various projections are in practice almost indistinguishable; and the simple conic is so satisfactory that it is not worth while to trouble about the more complicated forms which give equal area or orthomorphism.

For maps covering a larger extent of the sphere we shall find that it is very desirable to use two standard parallels instead of one. But so far as orthomorphism is concerned, the so-called orthomorphic projections are little better than the others at representing the true shapes of large configurations. As we have remarked before, orthomorphism is often more interesting mathematically than valuable practically, and so long as a map has no distortion in very small areas but considerable distortion in larger, so long will the fact that it is theoretically orthomorphic be of small value in practice. [K*]

The equal area property is of more practical importance, especially for statistical and political maps. And the reason why true conical equal area projections have not come into use may be found in the fact that there are two so-called "modified" conical projections which are equal area, and which are easier to draw than the true conical equal area projections. We shall discuss them later. (See Chap. VI, p. 52.)

We have already insisted that ease of drawing is a property which must be carefully considered. At first sight it may seem that all the true conical projections, built up of straight meridians diverging from a point, and circular parallels centred upon that point, must be easy to draw. This is true of very small scale maps; but when the scale is at all large the centre of the parallels is found to come at a considerable distance outside the map which is under construction. This introduces some difficulty in drawing the parallels, which can however be overcome. If space is available, it is not hard to construct a beam compass to describe arcs of radii up to say twelve feet. Or arcs of this large radius may be drawn by means of the circular curves which are often found in the equipment of a drawing office. But it is far more difficult to draw the set of straight meridians intersecting in a point twelve feet from the map, since this requires a very long straight edge, which is not so common.

There is however little difficulty in computing the positions of a pair of points on each meridian, from which it may be constructed; and there is not much excuse for the adoption of the so-called "simplified" conical projection, which has been much used in at least one modern atlas.

Oblique conical projections.

We have spoken, so far, only of *normal* conical projections—that is to say, of projections in which the vertex of the cone lies on the axis of rotation of the Earth, and the straight lines converging to the vertex represent meridians. There is, however, nothing in the nature of conical projections to confine them to this normal form. They may be constructed obliquely, so that the concentric circular parallels of the projection no longer represent parallels of latitude, but parallel small circles described about any selected point of the sphere. We shall have to refer to these projections at a later point; but we shall do so very briefly, since they are at present more curious than of practical importance. One or two maps have actually been constructed in this way in Germany, and small specimens of them are given in Zöppritz-Bludau*, p. 90, and Hammer†, plates II and IV. They have the remarkable defect that the meridians are not only not straight, but they have a sudden change of curvature in the neighbourhood of the principal axis of the map, which is a great disfigurement. It is difficult to believe that oblique conical projections will come into general use.

In all the true conical projections the angles which the meridians make with one another are controlled by a constant n which we have called the constant of the cone. The value of n lies between 0 and 1; the angle between any two meridians of the projection is n times the true angle between those meridians at the pole of the Earth; and consequently the whole map is comprised in a sector of which the angle at the vertex is $n.360°$.

Zenithal projections.

As n increases we may imagine the cone getting flatter and flatter until when n becomes equal to unity it becomes a plane, and the boundary of the whole map becomes a perfect circle, instead of a sector.

If it is a normal cone which is thus degenerating, it becomes

* Leitfaden der Kartenentwurfslehre...Karl Zöppritz ; herausgegeben von Alois Bludau, Leipzig 1899.

† Über die geographisch wichtigsten Kartenprojektionen .... E. Hammer, Stuttgart 1889.

eventually a plane touching the sphere at the pole. If the cone is not normal, it becomes a plane touching at some other point. And just as we have found it sometimes helpful to think of the conical projections as constructed on a cone which may be cut and rolled out flat, so we shall find it convenient to think of these degenerated conical projections as constructed upon a tangent plane to the sphere. They form an important class, the zenithal or azimuthal projections, in which the true bearings from the centre are preserved.

Cylindrical projections.

On the other hand, as the constant of the cone n becomes smaller, and the solid angle at the vertex of the cone more acute, we may imagine that the vertex is removed further and further from the sphere, until when n becomes zero its distance is infinite, and the cone has become a cylinder, touching the sphere along the equator if the axis of the cylinder is normal, and along some other great circle if the axis is oblique.

The cylindrical projections, with one exception, Mercator's, are not in common use. A few examples occur in German Atlases, to which we shall refer later. But, generally speaking, the normal cylindrical projections have little merit, while the transverse are complicated and difficult to draw without offering any noteworthy advantages. We may exclude them from lengthy consideration, for they can very obviously be derived from conical projections, if necessary. [L*]

Geometrically speaking, the zenithal projections are equally conical. But practically they form a class whose constructions, and whose merits, are so different from those of the conical projections, that it tends to clearness if we treat them separately. This follows more especially from the fact that while the use of conical and cylindrical projections is confined almost entirely to their normal forms, the zenithal projections are generally oblique, that is to say, the tangent plane on which we may imagine them constructed is not generally tangent at the pole, but at some other point of the sphere, in the centre of the proposed map. All of them preserve the true bearings or azimuths of all points from this centre, whence their name, azimuthal; and points which are equidistant from the chosen centre upon the Earth are found to be also equidistant, though not necessarily at the right distance, from the centre of the map. These are properties which differentiate them in practice from the usual conical and cylindrical projections, and will justify us in treating them as a separate class of projections.

Summary.

We may summarize our general consideration of the conical projections, then, as follows :

A conical projection should be defined as a projection in which a set of radiating great circles on the globe is represented by a set of straight lines radiating from a centre; and the corresponding system of small circles on the globe, at right angles to the great circles, by a set of circles described about this centre, and consequently at right angles to the radiating straight lines*.

This definition includes all cases of conical, cylindrical, and zenithal projections, oblique as well as normal.

But whereas conical and cylindrical projections are generally normal, zenithal projections are generally oblique. In the former case, the radiating straight lines are the representations of the meridians of longitude, the concentric circles of the parallels of latitude ; and this is so convenient and generally desirable that it tends in practice to separate this class of projections from the zenithal in which the radiating straight lines are lines of equal bearing from the centre, the concentric circles are lines of equal distance from the centre. Were these systems of lines shown upon the map, then the analogy with the ordinary conical projections would be evident. But in practice they are not shown. The lines which are shown are the meridians and parallels, and for oblique projections these are not straight lines and circles. Bearing in mind, therefore, the fact that all are included in the broad definition of conical projections, we shall nevertheless find it convenient to treat the zenithal or

---

* Colonel C. F. Close, C.M.G., R.E. *The Geographical Teacher*, Vol. IV, p. 158.

azimuthal projections apart from the ordinary conical and cylindrical.

There is also another reason for this division. Conical and cylindrical projections may be excellent for representing a portion, even a large portion, of the sphere. But they are very little use for the representation of a complete hemisphere, and still less of the whole sphere*. On the other hand, zenithal or azimuthal projections include some of the more interesting of those projections which will represent a hemisphere or more, and this tends to emphasize the practical convenience of separating zenithal from conical projections, even though from the purely geometrical point of view they ought to be kept together.

* It may appear that this statement is contradicted by the fact that Mercator's, a cylindrical projection, is much used for maps of the World. But this is done by sacrificing the polar regions. A map of a complete hemisphere on Mercator's projection extends to infinity.

CHAPTER III

CYLINDRICAL PROJECTIONS

WE have already remarked that cylindrical projections are limiting cases of conical projections, when the constant of the cone, n, becomes equal to zero. It will be convenient to go quickly through the list of conical projections which we have already treated, and see what happens to them when n becomes zero. Any of the resulting cylindrical projections which are of value can be considered more in detail afterwards.

Simple cylindrical.

In the simple conical projection with standard parallel of latitude ϕ_0 we have, with the usual notation,

$$n = \sin \phi_0, \quad r_0 = R \cot \phi_0, \quad \text{and} \quad r = R \{\cot \phi_0 - (\phi - \phi_0)\}.$$

If n is zero, ϕ_0 is zero, and the standard parallel is the equator. The radius r_0 of the equator on the projection becomes infinite; but $r_0 - r = R\phi$ and is finite.

We have as a result the simplest and most conventional of all projections, called by the French "projection plate carrée" and by the Germans "quadratische Plattkarte," but for which there does not seem to be any English name. The meridians and parallels are two sets of equidistant straight lines cutting at right angles and forming a series of squares. Distances along the meridians and along the equator are correct; distances along the other parallels very soon become glaringly incorrect; and the projection is of no value. We shall not consider it further. [M*]

Cylindrical with two standard parallels.

In the ordinary conical projection with two standard parallels of latitudes ϕ_1 and ϕ_2, we have, with the usual notation (see Chap. VIII),

$$n = \frac{\cos\phi_1 - \cos\phi_2}{\phi_2 - \phi_1}, \quad r_1 = \frac{R(\phi_2 - \phi_1)\cos\phi_1}{\cos\phi_1 - \cos\phi_2},$$

and

$$r_1 - r = R(\phi - \phi_1).$$

If n is zero, we have $\phi_1 = -\phi_2$, which makes r_1 infinite, but $r_1 - r$ remains finite.

We have as a result a very conventional projection differing from the last only in the respect that two parallels, at equal distances north and south of the equator, are represented correctly. The scale along parallels distant from the two standards is very incorrect. The meridians and parallels form a series of rectangles instead of a series of squares. Hence the French name "projection plate parallélogrammatique," and the German "rechteckige Plattkarte." There does not seem to be any English name for it, which is of small consequence, as the projection has obviously no serious value, and will not be considered further in this book.

Cylindrical equal area.

In the simple conical equal area projection, with one standard parallel of latitude ϕ_0 we have, with the usual notation,

$$n = \cos^2 \tfrac{1}{2}(90° - \phi_0), \quad r_0 = 2R\tan\tfrac{1}{2}(90° - \phi_0),$$

$$r = 2R\sec\tfrac{1}{2}(90° - \phi_0)\sin\tfrac{1}{2}(90° - \phi).$$

If $n = 0$, $\phi_0 = -90°$; that is to say, the south pole becomes the standard parallel. In this case the projection clearly shuts up into a line along the axis of the cone, normally the polar axis of the Earth, and is of geometrical interest only. We need not consider it further.

In Albers' conical equal area projection, with two standard parallels of latitudes ϕ_1 and ϕ_2, we have, with the usual notation,

$$n = \tfrac{1}{2}(\sin\phi_1 + \sin\phi_2) = 1/k, \quad r_1 = kR\cos\phi_1,$$

$$r^2 = 2kR^2(\sin\phi_1 - \sin\phi) + r_1^2.$$

When n is zero, $\phi_1 = -\phi_2$, that is, the standard parallels are equidistant north and south of the equator; r_1 becomes infinite, but we can show that $r_1 - r$ remains finite, and $= R \sin \phi$, the expression for the cylindrical equal area projection, which is of some value. The derivation of this projection from that of Albers' is interesting as showing how it fits into the general theory of conical and cylindrical projections. It is more usual, however, to derive it independently, as follows.

The cylindrical equal area projection is one of the few which are really projections in the geometrical sense. If we imagine a circumscribing cylinder touching the sphere along the equator, and if through any point of the sphere we draw a perpendicular to the axis, and produce it backwards to cut the cylinder, then it is evident from the figure that we have the projected point on the cylinder at a distance $R \sin \phi$ from the equator, which gives the same law for the projection that we found before. Also it is a well-known geometrical property that a small area on the sphere is unaltered by projection in this way on the cylinder. It follows that this geometrical projection of the sphere upon the circumscribing cylinder is an equal area projection.

Other properties of the projection are easily derived from the figure. Thus

(1) The scale along the parallels increases rapidly north and south of the equator, for all the parallels on the cylinder are equal, whereas the corresponding parallels upon the Earth are to the equator in the ratio $\cos \phi : 1$.

(2) The scale along the meridians decreases rapidly north and south of the equator, for in the projection a degree of latitude becomes more and more foreshortened.

(3) These considerations show that the projection is very far from being orthomorphic, although the meridians and parallels cut at right angles.

The projection is very easy to draw, but its great distortion and inequalities of scale in high latitudes make it of little use. We shall have no need to consider it in greater detail.

Cylindrical orthomorphic (Mercator).

In the conical orthomorphic projection with a standard

parallel of latitude ϕ_0, or co-latitude $\chi_0 = 90° - \phi_0$, we have, with the usual notation,

$$r = m (\tan \tfrac{1}{2}\chi)^n,$$

where
$$m = R\, \frac{\sin \chi_0}{n (\tan \tfrac{1}{2}\chi_0)^n},$$

and the value of n is still at our disposal. But if we wish the cone to have the same angle as the tangent cone along the parallel of χ_0, then $n = \cos \chi_0$.

If now we take the equator as the standard parallel, $n = \cos 90° = 0$; m becomes infinite, but $mn = R$ and is finite. Both r_1 and r become infinite.

We can however show, as will be done in Chap. VIII, that $r_0 - r$ is finite, and $= \dfrac{R}{M} \log \tan (45° + \tfrac{1}{2}\phi)$, where the logarithms are the common logarithms, and M is their modulus, whose reciprocal is 2·30259.

The parallels are then circles of infinite radii, or straight lines; and the distance of any parallel of latitude ϕ from the standard parallel, the equator, is given by

$$y = 2\text{·}30259\, R \log \tan (45° + \tfrac{1}{2}\phi),$$

the length of the equator, on the same scale, being

$$2\pi mn, \text{ or } 2\pi R.$$

Hence the cylindrical orthomorphic, or Mercator's projection, is a special case of the conical orthomorphic projection.

The meridians are parallel straight lines; and the distance of any one from the central meridian is given by $x = \dfrac{dL}{180°}\, . R$, where dL is the difference of longitude from the central meridian.

The parallels of latitude are parallel straight lines at right angles to the meridians, and we have just seen that their distances from the equator are given by

$$y = 2\text{·}30259\, R \log \tan (45° + \tfrac{1}{2}\phi).$$

The projection is orthomorphic, for it is a special case of the conical orthomorphic. And it is easy to prove the property independently, from the above expressions for x and y. (See p. 104.)

The scale along the parallels is evidently the scale along the equator $\times \sec \phi$; and rapidly becomes erroneous as the latitude increases.

Since the projection is orthomorphic, the scale along the meridian at any point is the same as the scale along the parallel. Hence the scale of areas at any point is the scale of areas at the equator $\times \sec^2 \phi$.

The distance between successive parallels increases rapidly with the latitude, and the poles are at infinity. Hence it is obvious that the projection is entirely unsuited for representing high latitudes.

The values of y are tabulated, and the projection is very easy to draw. It is usual to find in Atlases a map of the World on Mercator's projection, and these maps are responsible for very much misconception as to the size of northern countries such as Siberia, Greenland, or the northern portions of the Dominion of Canada. A comparison of the shape of Greenland on Mercator's projection and on a projection suitable for the polar regions respectively, gives a very excellent idea of how bad an orthomorphic projection may be in representing shape.

Use in navigation.

The great distortion in the north and south makes Mercator's projection altogether unsuitable for a land map. Its celebrity is due to the fact that it is invaluable for marine charts, and the reason is easily seen: the compass course between any two points on a Mercator chart is the straight line joining them. This follows at once from the two properties (1) that the projection is orthomorphic, and (2) that the meridians are parallel straight lines. Hence a ship's course from point to point can be taken at once from the chart.

We must be careful to distinguish, however, between the compass course and the great circle course. The straight line on a Mercator's chart is the compass course; that is to say, if the line drawn from A to B on the chart is 75° west of north, a ship starting from A and sailing continually on the course N. 75° W. arrives at B. (The variation of the compass is of course neglected in this statement.) The shortest distance from A to B is not the course of equal bearing throughout, but is the great circle course, which in the northern hemisphere will lie to the north, and in the southern hemisphere to the south of the compass course. We cannot enter into the question of great circle sailing here. (But see p. 41, on gnomonic or great circle charts.) The fact that a long voyage is not generally made on a uniform compass bearing does not alter the fact that the use of the Mercator chart in navigation is almost universal.

We shall consider its construction and properties more in detail in Chapter VIII.

CHAPTER IV

ZENITHAL PROJECTIONS

IN our General Remarks on the Conical Projections we mentioned very briefly the modification which a conical projection undergoes in the limiting case when the constant of the cone, n, becomes equal to unity. Suppose first of all that the projection is normal. The straight lines radiating from the vertex, which represent the meridians, then make the same angles with one another that the meridians do; they are no longer confined to a sector, but are disposed symmetrically round the vertex. The parallels are now complete circles, not arcs of circles, with the vertex as their common centre. And in the same way that it was found helpful sometimes to think of the conical projections as drawn upon a cone, so with the zenithal projections we may think of them as constructed upon a plane tangent to the sphere at the vertex.

We remarked further that the zenithal projections are not restricted in use to normal cases, as are the conical. The tangent plane is not necessarily tangent at the pole; but is usually oblique to the axis of the Earth, and tangent to the Earth at whatever point we may wish to take as the centre of our map. The equally spaced radii then represent, not of course meridians, but equally spaced great circles radiating from the chosen central point, lines of equal bearing or azimuth; while the concentric circles are no longer parallels of latitude, but represent circles of equal distance from the central point.

Following the plan which we adopted in our consideration of the cylindrical projections, we shall first consider briefly those zenithal projections which are only special cases of the conical projections with which we are already familiar.

Afterwards we shall consider some other zenithal projections, and among them the so-called perspective projections, which are not special cases of any conical projections in general use, though conical projections analogous to them could be constructed if it were desired.

Zenithal equidistant projection.

The Simple Conic with one standard parallel becomes the zenithal equidistant projection. It is clear by analogy that the standard parallel of the conical closes up into the centre of the zenithal projection, and that the parallel circles are spaced out at their true distances from the centre.

Hence in the zenithal (or azimuthal) equidistant projection the azimuths and the distances from the centre are true. The scale along the radii is everywhere true. The scale along the parallel circles is true only close to the standard parallel, that is to say, in this case true only close to the centre; at all other points the scale along the parallel circle is too large, and is increasingly erroneous as distance from the centre increases.

For example, if the radius of the sphere is R, the parallel circle 90° from the centre has radius $\frac{\pi}{2}R$, and its length is consequently $\pi^2 R$. But the true length of this circle on the sphere is $2\pi R$. Hence on the projection its length is $\frac{\pi}{2}$ or 1·57 times its true length.

It is clear, then, that the zenithal equidistant projection is far from being either equal area or orthomorphic. It is unsuitable for representing so large an area as a hemisphere, but is quite suitable for a map illustrating polar exploration, for example, and is frequently so used in Atlases.

The normal, or polar zenithal equidistant projection, is of course very easy to draw. It is equally easy to draw the radii and parallel circles for the oblique, or so-called horizontal projection, centred on some point of the sphere not the pole. But we generally wish to show on the map, not these radii of equal bearing and parallel circles of equal distance from the centre, but the meridians of longitude and the parallels of latitude, which are not simple curves. To construct these we may proceed in either of two ways:

I. We may construct them by transformation from another zenithal projection, the Stereographic, in which the curves in question are circles, and easily drawn geometrically. For this process, which is applicable to the whole group of zenithal projections, see Chapter VIII.

II. We may construct the meridians and parallels by calculating the azimuths and distances from the centre of a sufficient number of their points of intersection, plotting these points, and drawing curves through them. There are tables in existence in which a number of these azimuths and distances are already computed, and these will generally suffice. See Chap. X. We shall discuss this projection more fully in Chap. VIII.

It is evident that the ordinary conic with two standard parallels has no counterpart in the zenithal projections. For, as we have just seen, if the radii are divided truly, the scale along the parallel circles is too great everywhere except at the centre, and it is not possible to have two distinct parallels of their true lengths.

Zenithal equal area.

The conical equal area projection with one standard parallel becomes the zenithal equal area projection. In the normal case of the former, the radius of the parallel of co-latitude χ is given by

$$r = 2R \sec \tfrac{1}{2}\chi_0 \sin \tfrac{1}{2}\chi.$$

In the latter the radius of the parallel circle corresponding to an angular distance ζ from the centre is, by analogy, given by

$$r = 2R \sin \tfrac{1}{2}\zeta,$$

χ_0, the co-latitude of the standard parallel, corresponding to $\zeta_0 = 0$, the angular distance from the centre of the standard parallel circle.

In this projection azimuths from the centre are true, as in all zenithal projections. The scale along the parallel circles is too large; the scale along the radii is too small in inverse proportion, for the projection is equal area.

For example, the radius of the parallel circle representing an angular distance of 90° from the centre is, by the formula, $2R \sin 45° = \sqrt{2} \cdot R$, which is less than $\dfrac{\pi}{2} R$, the true distance. The circumference of the parallel circle is $2\pi \sqrt{2} \cdot R$, which is greater than $2\pi R$, the true length of the circumference.

Since the scale along the radii and the parallel circles is not the same, the projection is clearly not orthomorphic. We shall see, however, when we discuss the errors of scale in detail (see p. 106) that the distortion is not great up to 30° from the centre, and that the zenithal equal area projection is, in consequence, valuable, and much used in Atlases, for maps of a considerable area of the world, such as the continent of Africa, or Central Asia.

What we said on the subject of drawing the zenithal equidistant projection applies equally to the zenithal equal area.

It is clear that the conical equal area projection with two standard parallels (Albers') has no counterpart in the zenithal projections. The two standard parallels coalesce into the central point, and the general expression for the radius of the parallel of latitude ϕ, in the normal conical projection, viz.

$$r^2 = 2kR^2 (\sin \phi_1 - \sin \phi) + r_1^2,$$

becomes, when $\phi_1 = 90°$, $r_1 = 0$, $k = 1/n = 1$,

$$r^2 = 2R^2 (1 - \sin \phi)$$
$$= 2R^2 (1 - \cos \zeta),$$

which reduces to $r = 2R \sin \frac{1}{2}\zeta$, as above, and we have the ordinary zenithal equal area projection.

Zenithal orthomorphic (Stereographic).

The general expression for the Conical orthomorphic projection is

$$r = m (\tan \tfrac{1}{2}\chi)^n,$$

where χ is the co-latitude of a parallel, χ_0 of the standard parallel, and

$$m = \frac{R \sin \chi_0}{n (\tan \tfrac{1}{2}\chi_0)^n}.$$

Now when $n = 1$ and the standard parallel closes up into the central point of the map, we have by analogy for the zenithal orthomorphic projection

$$r = m \tan \tfrac{1}{2}\zeta,$$

and $\qquad m = \dfrac{R \sin \zeta_0}{\tan \tfrac{1}{2}\zeta_0} = 2R \cos^2 \tfrac{1}{2}\zeta_0 = 2R,$ since $\zeta_0 = 0.$

Hence the general expression for the radii becomes

$$r = 2R \tan \tfrac{1}{2}\zeta.$$

The zenithal orthomorphic projection is usually called the Stereographic. It is of great interest theoretically because it is common ground of three different groups of projections. We have just seen that it is a particular case of the conical orthomorphic projection. It is also the most elegant and useful of the Perspective zenithal projections, the one which is easiest to draw, and from which the others may be obtained by transformation. Finally it is a particular case of the group of projections. very interesting mathematically, but of no practical use, which are included under the general name Lagrange's Circular Orthomorphic projection.

We shall defer the consideration of this projection until we treat it in its proper place among the perspective projections.

We have seen in this and the preceding chapter how the zenithal and cylindrical projections are related to the conical. It will be convenient to make a small table showing the relationships.

Table of Related Projections.

	Conical	Cylindrical	Zenithal
1.	Simple conic with one standard parallel	Simple cylindrical " plate carrée "	Zenithal equidistant
2.	Conical with two standard parallels	" Plate parallélogrammatique "	————
3.	Conical equal area with one standard parallel	————	Zenithal equal area
4.	Conical equal area with two standard parallels	Cylindrical equal area	————
5.	Conical orthomorphic	Cylindrical orthomorphic or Mercator's	Zenithal orthomorphic or Stereographic

Airy's zenithal projection " by balance of errors."

We have seen that the expression

$$r = 2R \sin \frac{\zeta}{2}$$

gives the zenithal equal area projection, in which the shapes of all areas, even very small ones, are distorted, and sacrificed to the preservation of areas; while the expression

$$r = 2R \tan \frac{\zeta}{2}$$

gives the zenithal orthomorphic, in which the shapes of small areas are preserved, but are subject to great errors of scale.

Sir George Airy conceived the idea of making a zenithal projection which should be a kind of happy mean between these two extremes.

In the first the scale along the radii is too small and the scale along the parallel circles too great in inverse proportion at any point. In the second the scales along the radius and the parallel circle at any point are the same, but they are too large everywhere except at the centre.

Airy argued that the "total misrepresentation" of the map might be expressed by the sum of the squares of the errors of scale in the two directions taken for every point of the map; and he determined the law of a projection which made this sum a minimum.

The law evidently depends upon the extent of the spherical surface to be represented, for the larger the area to be shown on the map, the greater is the difficulty of representing the outlying portions without undue distortion, and the greater the sacrifice which has to be made in the representation of the central regions to allow some mitigation of the distortion at the edges.

By a rather complex theoretical investigation Airy deduced the following law :

If β is the spherical radius of the portion of the sphere to be represented, then with the usual notation

$$r = \frac{2R}{M} \{\cot \tfrac{1}{2}\zeta \log \sec \tfrac{1}{2}\zeta + \tan \tfrac{1}{2}\zeta \cot^2 \tfrac{1}{2}\beta \log \sec \tfrac{1}{2}\beta\},$$

where the logarithms are common logarithms, and M is their modulus, whose reciprocal is 2·30259.

We have seen that the projection is, by intention, neither orthomorphic nor equal area, being designed as some sort of a mean between the two.

It is difficult to give an idea of the numerical properties of this projection in a small compass, as they vary with the radius adopted for the boundary. The only example of the projection to be found in Atlases is drawn for the spherical radius $113\frac{1}{2}°$, with centre in Lat. $23\frac{1}{4}°$ North and Long. $15°$ East. This includes nearly the whole land surface of the globe. At $90°$ from the centre of the map the radius is only about $8°/_0$ greater than its true length; areas are exaggerated $2·2$ times, and the ratio of the scale along the parallel circle to the scale along the radius is only $1·30$. At $110°$ from the centre these quantities become $14°/_0$; $3·97$; and $1·39$ respectively. From this it is clear that Airy's projection "by balance of errors" is remarkably successful in representing enormously large areas without excessive distortion. [N*]

It is a little complicated to compute; but the labour of computation is in any case only an insignificant fraction of the whole labour of producing a map, and need hardly be taken into account. When once the computation is done, the drawing is as easy as the drawing of any zenithal projection, and may be conducted either by transformation from the stereographic or by the aid of tables. See p. 110.

The projection may be recommended strongly for the representation of a hemisphere, but has not, to the knowledge of the author, been used for this purpose in any Atlas.

The Ordnance Survey map of the United Kingdom on the scale of 10 miles to the inch is constructed on this projection, but the whole area represented is too small to make a fair example of its remarkable merits.

CHAPTER V

ZENITHAL PROJECTIONS, CONTINUED

Perspective projections.

WE have now to consider a class of zenithal projections which, in a sense, stand by themselves, because they include all but one or two of the map projections which are projections in the strict geometrical sense of the word, made by projecting a portion of the sphere upon a plane by straight rays proceeding from a point, the centre of projection. The image formed on the plate of a "pin-hole" camera is an excellent example of a perspective projection.

The properties of perspective projections naturally depend upon the position of the centre of projection.

Imagine a diameter of the sphere, and a tangent plane to the sphere drawn at one extremity. And let us consider briefly the projections which can be made from different centres of projection lying upon the diameter. [O*]

It is evident that all such perspective projections are zenithal or azimuthal, that is to say, that any set of great circles of the sphere radiating from the point of contact of the plane and the sphere, the centre of the map, project into radial straight lines making the same angles with one another as do the great circles. And small circles of the sphere, described about the point of contact, project into parallel circles described about the centre of the map.

I. If we project from the centre of the sphere we have the Gnomonic projection.

II. If we project from the opposite end of the diameter we have the Stereographic projection.

III. If we project from a point on the diameter produced, outside the sphere on the side opposite to the tangent plane, we have a series of projections comprised in the general name of Clarke's Minimum Error Perspective projection. In the useful cases the centre of projection lies at a distance from the centre of the sphere between 1·65 and 1·35 times the radius of the sphere.

IV. If we choose the particular case of the last in which the distance is 1·367 radii we have Sir Henry James' projection.

V. If we choose the particular case in which the distance is $1 + \dfrac{1}{\sqrt{2}}$, or 1·71 radii, we have La Hire's projection.

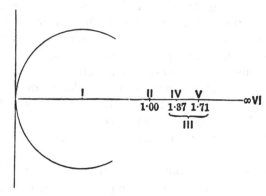

Fig. 5.

VI. Finally, if we take the centre of projection at infinity, or project by straight lines perpendicular to the tangent plane, we have the orthographic projection.

The gnomonic projection.

If ζ be the angular distance of any point on the sphere from the point which is to be at the centre of the map, it is clear that the linear distance from the centre of the projected point is given by

$$r = R \tan \zeta.$$

As in all zenithal projections, the true bearings from the centre are preserved in the projection. But the distances from the centres are very much distorted, since tan ζ increases much more rapidly than ζ.

For example if $\zeta = 45°$, $r = R$, whereas the true length on the sphere of an arc of $45°$ is $\frac{\pi}{4}R = 0\cdot785R$.

As the distance from the centre increases, the scale along the radii becomes rapidly greater, and as ζ approaches $90°$ it increases without limit, as is evident from the fact that

$$\tan 90° = \infty ,$$

or geometrically, from the fact that when $\zeta = 90°$ the projecting rays are parallel to the plane which they are required to cut.

The scale along the parallel circles is also too great except very close to the centre. For $\zeta = 45°$ the circumference of the parallel circle is $2\pi R$; while the circumference of the corresponding small circle upon the sphere is $2\pi R/\sqrt{2}$. And as ζ increases the error becomes rapidly greater.

It follows that all distances, areas, and shapes are represented very badly on the gnomonic projection, which would be quite useless if it had not one very valuable property—that *any* great circle on the sphere is represented by a straight line upon the map.

The proof of this proposition is extremely simple. The plane of any great circle upon the sphere passes through the centre of the sphere. Hence the rays which project the great circle lie in one plane, and the gnomonic projection of the great circle upon any plane is a straight line.

The determination of the great circle course from point to point is an important problem in navigation. On a gnomonic chart the great circle course is the straight line joining the two points, and it would appear at first sight that such charts would be of great use to navigators. The United States Hydrographical Department has published a few charts upon this projection, but they do not seem to have come into general use. The reason is probably this, that the theoretical advantages of such charts are discounted in practice by the following facts: that great circle courses are required for very long voyages, whereas the distortion of the gnomonic chart is so great that it is not possible to represent anything like a hemisphere upon it with advantage: that strict great circle courses are often impracticable, owing to their taking the ship too far north or south; for example, the actual course from New York to Queenstown has to follow a parallel for some distance to avoid the regions of icebergs, and then turns on to the great circle: and finally, that the great circle courses along the principal routes are well known and laid down, or may be computed very readily with tables such as Lecky's.

Hence charts upon the gnomonic projection are very little used, and maps on it are hardly ever found in Atlases.

Projection of the sphere on the circumscribed cube.

The particular case of the gnomonic projection of the whole sphere upon the enveloping cube has some points of interest. If we regard a map of the world as, in the first place, a guide to getting about the world by the shortest route, then all the projections usually found in Atlases are quite inadequate. There is not one which will give with any facility the answer to such a question as: What course will a ship take from the Panama Canal to Yokohama? or, In what direction would an aviator start from London, to go straight to Sydney?

Such questions can be solved very readily by the gnomonic projection on the cube. (See Fig. 9, p. 45.)

Consider first the case in which the cube touches the sphere at the poles.

The construction of the two polar sides is very easy. The parallels of latitude are circles, with radii $R \cot (\text{latitude})$; and the meridians are equally spaced straight lines radiating from the poles.

Consider next the equatorial sides of the cube. The equator is a straight line bisecting these four sides. The meridians are straight lines perpendicular to the equator, and distant from the central meridian $R \tan (\text{diff. of long.})$. The parallels are hyperbolas. A parallel of latitude ϕ cuts a meridian distant ΔL from the central meridian at a point whose distance from the equator is $R \tan \phi \sec \Delta L$. Hence it is very simple to construct the points where the parallels intersect the meridians, and to draw the corresponding curves through them.

Suppose next that the cube does not touch the sphere at the poles, but that its points of contact have been chosen to

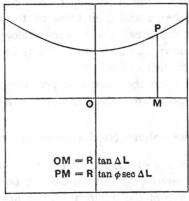

$$OM = R \tan \Delta L$$
$$PM = R \tan \phi \sec \Delta L$$

Fig. 6.

suit the requirements of the cartographer, with a view to arranging the land surfaces most conveniently upon the sides of the cube. We must obtain a few elementary properties of the projection.

Construction of a great circle.

The fundamental problem is to draw the great circle joining any two points. If they are on the same face of the cube, the great circle is of course the straight line joining them.

This great circle may be continued on an adjacent face very simply. Suppose two contiguous faces laid out flat, and let L, M, N be the middle points of the three consecutive parallel edges. (We should note that for the purposes of construction the faces need not be considered as limited by the edges of the cube, but may be produced indefinitely. But to make this clear, we shall show dotted those lines which are drawn upon such extensions of the faces. See figure on following page.)

Let AB be a great circle on one face. If BC is its continuation on an adjacent face, we find C from the consideration that, by symmetry, C must be as much above N as A is below L.

Hence we have the rule for joining two points P, Q on adjacent faces: Find by trial a point S on the common edge, such that the lines SP, SQ make equal intercepts respectively above and below the middle points of the adjacent parallel edges. This is easily done by trial; but it does not appear that there is any direct construction that will give the point S in a simple way. [P*]

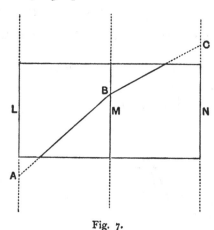

Fig. 7.

If the points P and Q are on opposite faces the solution is still more simple. Find the point P' opposite P, or Q' opposite Q, and join $P'Q$ or $Q'P$. This gives part of the required great circle, which may be continued round the cube in the way just explained*.

The oblique cube.

We are now in the position to deal with the case of the cube which does not touch at the poles.

Let the centre C of one face be in latitude ϕ, and let the meridian through C be drawn parallel to sides of the square. The equator LM is perpendicular to this meridian, at distance $R \tan \phi$, and the poles N, S are the same distance above or below the centres of their faces.

The equator may be continued round the other faces by

* A good discussion of the geometry of this projection will be found in a paper by Professor Turner, *Monthly Notices of the Royal Astronomical Society*, Vol. LXX, p. 204.

the rule given above; it is *KLMNO* in the figure. *KL* and *MN* pass through the centres of their faces. The meridians which cut *KL* and *MN* are perpendicular to the equator, and may be constructed as in the normal case already considered; they may then be continued across the adjacent faces, to pass through *N* and *S*. The places where the meridians cut *LM* and *NO* may be found from the formula:

distance from centre of *LM* or *NO* $= R \tan \Delta L \sec \phi.$

And these points may be joined to the poles by the rules given above.

Thus all the meridians are constructed.

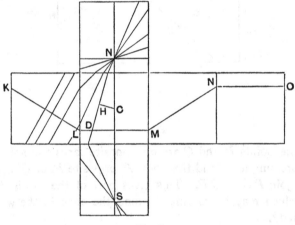

Fig. 8.

The parallels on the faces *KL* and *MN* may be constructed as in the normal case, or traced from it.

The parallels on the other faces are somewhat more troublesome.

From *C* draw a perpendicular *CH* to any meridian, and let $\theta = \tan^{-1} CH/R$.

Then the latitude of *H* is $\tan^{-1}(\cos \theta . DH/R)$.

Call this ϕ_0.

The distance of any parallel of latitude ϕ from *H* is then $R \tan (\phi - \phi_0) \sec \theta$, as before.

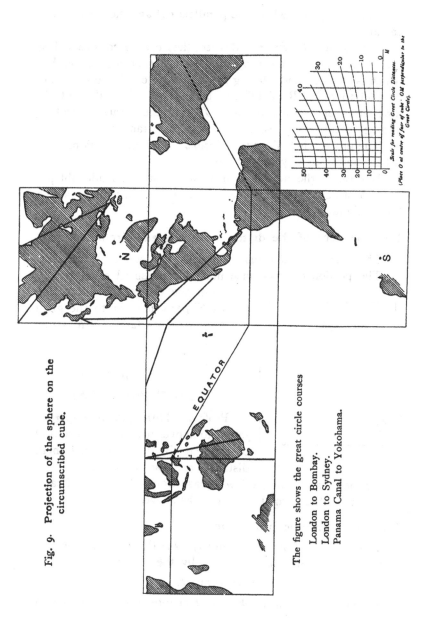

Fig. 9. Projection of the sphere on the circumscribed cube.

The figure shows the great circle courses

London to Bombay.
London to Sydney.
Panama Canal to Yokohama.

Thus the points where the parallels cut any meridian can be calculated.

If great accuracy is not desired, they can be constructed by taking a tracing of the projection on a normal face, or on *KL*, which is similar; and placing it with centre on *C* and meridians parallel to *HD*. Read off the latitude of *H*. Take out the differences of latitude between *H* and the desired parallels; and set them off by estimation from the tracing.

The process is tedious to describe. With a little familiarity it becomes quite easy in practice.

If desired, the meridian through *C* may be made oblique. The construction then proceeds in a way similar to the above, but is more tedious.

Measurement of the distance on a great circle.

The method is obvious from what immediately precedes. The portion on each face must be measured separately.

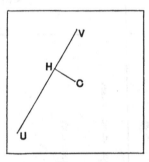

Fig. 10.

From the centre *C* of any face draw *CH* perpendicular to the great circle, and let $\theta = \tan^{-1} CH/R$, as before. Then

$$UV = UH \pm HV$$
$$= \tan^{-1}(\cos\theta \,.\, UH/R)$$
$$\pm \tan^{-1}(\cos\theta \,.\, VH/R).$$

Professor Turner, in the paper referred to above, has described a kind of protractor by means of which these distances are readily computed.

Or they may be read off with ease from a tracing of a carefully constructed normal projection, as above.

The Stereographic projection.

It is evident from the figure (p. 39) that the radii in the stereographic projection are given by the formula

$$r = 2R \tan \tfrac{1}{2}\zeta.$$

We shall prove in a subsequent chapter that the projection is orthomorphic. (See p. 100.)

The scale along the radii is everywhere too large, but it increases very much more slowly than in the case of the gnomonic projection, and at least a hemisphere can be represented without extraordinary distortion. For example, the radius of the parallel circle bounding a hemisphere is evidently $2R$, while the true distance is $\frac{1}{2}\pi R = 1·57R$.

Similarly, the scale along the parallel circles is everywhere too great, but it does not increase violently with distance from the centre. For example, the length of the circumference bounding the representation of a hemisphere is evidently $4\pi R$, while its true length is $2\pi R$.

It follows that the areas at a distance from the centre are considerably exaggerated, but not out of all proportion, as is the case with the gnomonic projection.

With regard to the representation of shape : since the projection is orthomorphic, very small areas are represented of their true shape. But since the meridians are curved, being circles, the shape of large areas is not well preserved.

The great merit of the stereographic projection is that it is easy to construct geometrically. The meridians and parallels are all circles, and it is easy for example to draw the stereographic projection of a hemisphere upon a small scale. When, however, we require maps upon a larger scale, representing a smaller portion of the Earth, the advantages of the geometrical construction disappear; it wants an impossibly large sheet for its construction. In practice, therefore, it offers no advantages over, let us say, the conical orthomorphic projection, for Atlas maps of a large, but not excessively large country, and its use appears to be decreasing.

Clarke's minimum error perspective projection.

Sir George Airy's principle of making the "total misrepresentation" a minimum has been applied to the perspective projections by Colonel A. R. Clarke, R.E. (Airy's projection is not perspective.)

It is easily seen that if a perspective projection be made from an external point, whose distance from the centre of the sphere is hR, R being the radius of the sphere, then the radii of the projection upon the tangent plane are given by

$$r = R(1 + h)\frac{\sin \zeta}{h + \cos \zeta}. \quad [Q^*]$$

The value to be chosen for h naturally varies with the extent of the map for which the total misrepresentation is to be a minimum. It does not seem possible to obtain an expression from which h can be calculated directly for any given spherical radius β which is to be the bounding radius of the area to be projected. The determination of h has to be made by a process of trial and error. Colonel Clarke has shown that for a map of Africa or South America, which can be included in a small circle of the sphere of radius 40°, $h = 1\cdot625$. For a map of Asia, with radius 54°, $h = 1\cdot61$; for the hemisphere, $h = 1\cdot47$; and for the radius $113\frac{1}{2}°$, $h = 1\cdot367$.

Two specimens of these maps, on very small scales, are given in Colonel Clarke's article on Mathematical Geography in the *Encyclopaedia Britannica*. The projection of the hemisphere ($h = 1\cdot47$) is decidedly better than the Stereographic as a projection of generally good qualities; but it has not, to the knowledge of the author, been used in Atlases.

The projections are easy to compute, when once the value of h has been decided, and they can be drawn either by transformation from the Stereographic, or by means of the table of azimuths. (See p. 110.)

Sir Henry James' projection.

This is simply Clarke's projection for the special case

$$\beta = 113\frac{1}{2}°, \quad h = 1\cdot367.$$

It has been drawn, so far as is known to the author, for one case only. The centre of the map is taken as the point where the meridian 15° East of Greenwich cuts the Northern Tropic. The South Pole is then on the circumference of the map, and the central meridian extends upwards across the North Pole, and 47° beyond. The reason for the choice of these coordinates appears to be that nearly the whole of the land surface of the world is then included within the circular boundary of the map.

An example of it may be seen in Philip's Systematic Atlas, plate 7. On the same plate is a map of the same area drawn on Airy's projection by Balance of Errors. On the small scale upon which they are drawn the two maps look exceedingly similar. A numerical examination shows, however, that Sir Henry James' projection represents radial distances slightly better, areas very much better, and shapes decidedly worse, than does Airy's projection. Both are, however, better than the stereographic projection in all respects, even the latter of the three just mentioned, the representation of shapes. For though the stereographic is truly orthomorphic, that is, represents small areas of their true shape, yet its rapidly increasing exaggeration of areas, more than twice that of the other two at a distance 90° from the centre, makes its representation of large configurations decidedly inferior.

Both Airy's and James' projections are based on the principle of making the sum of the squares of the errors of scale in two directions, taken all over the map, a minimum, and the results, for the particular case of spherical radius $113\frac{1}{2}°$, look so very much alike that the distinction between them is at first sight not always obvious. It may however be stated concisely as follows: Airy's projection is the solution of the problem, to construct a zenithal projection in which the total misrepresentation shall be a minimum; James' projection results from the solution of a slightly more restricted problem—the projection must be of the special class of zenithal projections which are perspective, true geometrical projections.

It is hardly going too far to say that Clarke's Minimum Error perspective projection, of which James' is a particular case, is the only true geometrical projection which is really good for the representation of a hemisphere or larger portion of the sphere.

The orthographic projection.

The orthographic is the ultimate case of perspective projection, where the centre of projection is removed to infinity, whence the projecting rays become parallel, and all perpendicular to the tangent plane upon which the projection is supposed made.

As a map projection for geographical purposes the ortho-graphic has no merits whatever, and will not be considered further in this portion of the book. It has however considerable interest to astronomers. Our world as seen from the great distances of the heavenly bodies appears orthographically pro-jected ; and maps of the hemisphere visible from a given direction at a given instant have been found very useful in studying the complex circumstances connected with such phenomena as the Transit of Venus across the disc of the Sun. For an admirable use of such diagrams reference may be made to the work of the late R. A. Proctor: *The Universe and the Coming Transits.*

CHAPTER VI

THE MODIFIED CONICAL AND CONVENTIONAL PROJECTIONS

THE deficiencies of the simple conic projection in the representation of areas have led to an extensive use of the so-called modified conical projection of Bonne, which is not in reality a conical projection at all, since the meridians are not represented by straight lines, even in the normal projection. The Sanson-Flamsteed projection is a particular case of Bonne's.

Another modification of the conical projection, the polyconic, has resulted from the circumstance that it is convenient, if it is possible, to compute once for all a set of tables which will enable a draughtsman to construct at once, without any preliminary calculation, a projection for any desired map. The Polyconic Projection possesses this advantage, and it is in consequence very largely used in some surveys, notably the United States Coast and Geodetic Survey. The mechanical ease with which it relieves the draughtsman of all responsibility for the choice of projection is its chief title to consideration, for it possesses no particularly valuable property such as the equal area property of Bonne's projection. It has, moreover, the defect that the meridians do not cut the parallels at right angles.

To avoid this defect, the Rectangular Polyconic was devised, and has been used extensively in the maps of the Intelligence Department of the British War Office, now the Geographical Section of the General Staff.

Neither the ordinary nor the rectangular polyconic is suitable for maps of a large area; and for small areas they are indistinguishable.

We may consider these successive modifications of the true conical projections as links between those projections which have some definite scientific value, and the projections generally called conventional which have no scientific interest, but possess the valuable properties of convenience and simplicity in use.

The Projection by Rectangular Coordinates, adopted for all the maps of the Ordnance Survey of England (except some of the smaller scales) is a step further towards complete conventionality. [R*]

And at the end of the series we have the purely conventional projections such as the globular, of some use for unimportant atlas maps, and the most easily drawn projection for a hemisphere, but of no other interest whatever.

We will consider these briefly in order.

Bonne's projection.

The scale of the simple, conical projection is correct along all the meridians, and along one selected parallel. The scale along the other parallels is incorrect, and the error becomes large if the projection is used for a map having a considerable extension in latitude. Bonne's projection remedies this defect in the following way.

The central meridian is drawn straight and divided truly; the parallels at their true distances apart are drawn as concentric circles; and the selected standard parallel is divided truly, all as in the simple conical projection. But the meridians are no longer formed by joining the vertex to the points of division of the standard parallel. Instead of this all the parallels are divided truly, and the meridians are formed by drawing curves through the corresponding points on each parallel. The resulting curves are not simple, but since a series of points upon them is so easily constructed, they are not hard to draw.

It is easily shown, and is indeed obvious geometrically, that the projection is strictly equal area. This has given it its popularity.

The scale along the parallels is correct everywhere, by construction. The scale along the meridians is not correct, as is obvious from the fact that the distances between the parallels are correct, but the meridians do not cut the parallels at right angles. The scale along the meridians is consequently too great for all except the central meridian, and this defect becomes more and more pronounced as the difference of longitude from the central meridian increases, and the meridians become more and more inclined to the parallels.

The considerable inclination of the meridians to the parallels away from the central meridian shows that the projection is very far from being orthomorphic. The same fact shows that it does not possess the more general quality of preserving general shapes fairly well in spite of the want of strict orthomorphism.

It is easy to draw when the proper appliances are at hand for passing curves through a series of plotted points.

It is clearly unsuitable for polar regions, as are all the conics, and the less its extension in longitude the better.

It was used for the old general map of France, whence its continental name of "projection du dépôt de la guerre." It is also used by many other European surveys, including the Ordnance Surveys of Scotland and Ireland; and it is very common in Atlases.

Sanson-Flamsteed projection. [S*]

This is the particular case of Bonne's where the equator is chosen for the standard parallel. Its properties are therefore the same as that of Bonne's, and there is no need to consider it separately.

It is often used in Atlases for the general map of Africa, whose extent in latitude is divided nearly equally by the equator, and whose extent in longitude is not great. It is also used for a general map of Australia and Polynesia, for which it is not well suited, since the extent in longitude is too great.

Werner's projection.

This is the particular case of Bonne's projection in which the standard parallel is at the pole, the cone then becoming the tangent plane at the pole. Any one meridian is chosen as the central meridian, and is a straight line truly divided. The parallels are divided truly, and the other meridians are curves, as in Bonne. It does not appear that the projection has any properties specially valuable; but it has been used in Schrader's *Atlas de Géographie Historique* for a map of the Russian Empire.

Polyconic projection.

This projection owes its name to the fact that *each* parallel is constructed as if it were the chosen standard parallel for an ordinary simple conic projection.

The central meridian is divided truly. The parallel of latitude ϕ is a circle of radius $r \cot \phi$, whose centre lies on the central meridian, and which cuts it at the proper point of division. Each parallel so constructed is divided truly; and the meridians are formed by passing curves through the corresponding points on successive parallels, as in Bonne's projection.

The parallels are not concentric circles, for the distance of the centre of each, measured along the central meridian, from the parallel of latitude ϕ_0 is evidently $r(\cot \phi + \phi - \phi_0)$ which decreases as ϕ increases. Hence the parallels diverge from one another on each side of the central meridian. This in itself would make the scale along the meridians wrong. The error is aggravated by the fact that the meridians do not cut the parallels at right angles.

It is clear that the projection is far from being either equal area or orthomorphic, and it is therefore quite unsuitable for maps of large area, for which indeed it is never used. Its value lies in the fact that a general table can be calculated for the polyconic which depends only on the adopted values for the size and shape of the Earth. For the radius of each parallel depends only on its latitude, and not in the least upon the position of the centre or the extent of the map. The meridians are not divided truly, but all meridians at equal distances from the central meridian are divided similarly. The parallels are divided truly.

It follows from these properties that if we have a map divided into a number of sheets, each covering the same extent of longitude, adjoining sheets will fit exactly along their northern and southern boundaries, for the bounding parallels are of definite radii and truly divided; and they will have a rolling fit along the eastern and western boundaries, for meridians at equal distances from the centre are divided similarly but are curved in opposite directions. On maps of the usual size and scale of a topographical map, this curvature of the meridians is hardly noticeable, and thus it is very nearly true that any two adjacent sheets plotted separately on the polyconic projection, fit one another at their junctions. This is a valuable property for a topographical map.

We have already remarked that the polyconic projection is not suitable for an atlas map. The sixth International Geographical Congress (London 1895) recommended the polyconic projection for the one in a million map of the whole world whose production was then advocated. The International Map Committee of 1909 finally adopted a slightly modified polyconic projection (see p. 56).

For the topographical maps of a country this projection has great conveniences:

(1) Each sheet may be plotted independently, but without any special calculation, by the aid of tables, constructed once for all. Excellent tables for the special scales 1/1,000,000, 1/250,000 and 1/125,000 are published by the British War Office as appendices to the official Text Book of Topographical Surveying. Specimens of these are given at the end of the book. The best general tables are those published by the United States Coast and Geodetic Survey.

(2) Adjacent sheets practically fit, as we have already remarked.

But they do not really fit, and it would not be possible to make a very large wall map by piecing together many such sheets; nor to make a small scale map of the country by photographically reducing such a pieced-together map. There are certain disadvantages, then, attached to the polyconic projection when used for a topographical series; and for a country not too large the projection by rectangular coordinates, on which the English Ordnance Survey maps are made, is in this respect preferable (see pp. 59 and 66). The question of fit is really only of importance in combining original plates or transfers on stone and zinc. The printed sheets suffer so much deformation by stretch of the paper that they can never be fitted together precisely in practice, even if they fit theoretically.

The rectangular polyconic.

This is also known as the War Office projection, because it has been very much used for the maps published by the Intelligence Department of the British War Office.

The parallels are constructed as in the polyconic projection. But they are not all divided truly. A selected parallel is divided truly, and the meridians are curves through the points of division of this standard parallel, which cut the other parallels at right angles.

The points of intersection of the meridians and parallels may be found by means of the tables published by the War Office*. They may also be constructed geometrically.

The simple and the rectangular polyconics differ very much from one another if they are drawn for the whole sphere, as in Germain, plate XIII. But they are never used for anything larger than a single sheet of a topographical map, and for this they are indistinguishable one from the other. There is, then, no practical need to consider the difference between the two; and as neither has any particular scientific interest, such as equal area or orthomorphic properties, we need not consider their theory any further. [T*]

We shall give examples of the construction of these projections in Chapter X.

The projection for the International Map on the scale of
1 : 1,000,000.

In general principle this projection is polyconic; but some interesting modifications are introduced.

The sheets cover four degrees of latitude and six of longitude (or in latitudes higher than 60° twelve degrees of longitude).

The top and bottom parallels are constructed from tables, in the usual way (see below). They are divided truly. They are, of course, circles struck from centres lying on the central meridian, but not concentric. The meridians are straight lines

---

* *On the Projection of Maps*, Major Leonard Darwin, R.E. 1890.

joining the corresponding points of the top and bottom parallels. Any sheet will now join exactly along the margins with its four neighbours. This is the first modification; in the true polyconic projection the meridians (except the central) are curves.

In the true polyconic projection the central meridian is divided truly by the parallels, and the other meridians are too long, by small amounts increasing with distance from the central meridian. In the projection for the International Map a second modification is made, in bringing the parallels closer together by a small amount, so that the meridians two degrees on each side of the centre are made to be of their true length. The correction required to effect this is very slight, amounting to only 0·01 in. at the most; and the consequent gain is therefore almost too small to be measured on the sheet. But the idea is elegant, and will appeal to all interested in the subject.

In the Resolutions of the International Map Committee, London, 1909, it is not laid down how the meridians are to be divided; but it may be supposed that they are divided equally. Nor is it provided that in sheets covering twelve degrees of longitude, instead of six, the meridians of true length shall be four degrees, instead of two, on each side of the central meridian; but this is no doubt intended.

The tables annexed to the Resolutions of the Committee are extracted from " Tables for the projection of graticules for maps on the scale of 1 : 1,000,000; prepared by the Geographical Section of the General Staff. London, Feb. 1910." But it should be noticed that the latter, though published immediately after the meeting of the International Committee, were not intended to include the above modifications. Thus the parallels for every degree are constructed separately and the meridians are not reduced to straight lines equally divided.

These tables provide for the construction of the parallels thus: Taking the point where the parallel cuts the central meridian as origin; the central meridian as the axis of y; and a line at right angles to it as the axis of x: the coordinates x, y are tabulated for the points of the parallel at longitudes 1°, 2°, and 3° from the central meridian. The y-coordinates are of course small. Thus seven points on each parallel are plotted, and a smooth curve passed through them is the required circular parallel.

If v is the radius of curvature of the spheroid at right angles to the meridian, ϕ the latitude of the parallel, and ΔL the difference of longitude from the central meridian,

$$x = v \cot \phi \sin (\Delta L \sin \phi),$$

$$y = v \cot \phi \{1 - \cos (\Delta L \sin \phi)\}.$$

[It may be noticed that in the above War Office tables published Feb. 1910, the values of x have, by an oversight, been calculated from the erroneous formula

$$x = v \cot \phi \tan (\Delta L \sin \phi).$$

The resulting error in the tables is very small, and practically almost negligible. It has unfortunately been reproduced in the International Map tables.] [U*]

The complete tables for the construction of the whole of the International Map up to 60° latitude are given on two pages (see pp. 114, 115) and no further computation is required for any sheet. This shows very clearly the practical advantages of the polyconic projection and its modifications.

An interesting discussion of the numerical properties of the International Map projection is given by M. Ch. Lallemand in the *Comptes Rendus*, Tome 153, p. 559. He finds that the maximum error of scale of a meridian is 1/1270, which corresponds to 0·35 mm. in the height, 0·44 m., of the sheet. The maximum error of scale of a parallel is 1/3200. And the greatest alteration of azimuth is six minutes of arc. These errors are much smaller than those occasioned by the expansion and contraction of the sheet by damp.

The plane table graticule, or field rectilinear projection.

This is nothing more than an approximation to the previous polyconic projections, in which the points of intersection of the meridians and parallels are joined up, not by curves, but by straight lines. The reason for doing so is obvious. When the trigonometrical triangulation is finished the plane tabler has given him a list of the geographical coordinates, latitude and longitude, of the triangulation stations and the intersected points. These he has to plot upon his plane table. The best way ot doing this is to begin by constructing the "graticule" or network of meridians and parallels. The plane table sheet covers a very small area, generally less than a quarter of a degree square. Within this area the curvature of the meridians and parallels is extremely slight. And the graticule has to be plotted in camp, away from the facilities of the drawing office. Hence it is an obvious simplification to draw the parallels and meridians in sections of straight lines, instead of as continuous curves. And the difference is quite inappreciable on the plane table sheet.

This projection has no properties of interest, except the all important one that it is usable in the field.

Excellent tables for the construction on various scales are published by the Survey of India (Auxiliary Tables) and in the *Text Book of Topographical Surveying* (Close). An extract from these tables will be found in Chapter X, with explanation of the method of use.

Projection by rectangular coordinates.

This is a step further towards the purely conventional projection.

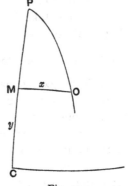

Let C be the centre of our map, of latitude ϕ_0, longitude λ_0; and let O be any other point, of latitude ϕ, longitude λ.

Draw the great circle OM perpendicular to the meridian through C, and let $OM = x$, $MC = y$.

Then it is easy to prove that

$$\sin x = \sin (\lambda - \lambda_0) \cos \phi$$

and

$$\cot (\phi_0 + y) = \cos (\lambda - \lambda_0) \cot \phi.$$

Fig. 11.

If now, having computed by these expressions the lengths of the arcs x and y, we plot them as *rectangular coordinates on a plane*, we have the projection by rectangular coordinates, or Cassini's projection.

This is the projection upon which Cassini made his great map of France. It was adopted for the one-inch map of England, and for the six-inch map of the United Kingdom; and for several continental surveys.

It is more conventional than the polyconic projection, and differs from it in the important respect that the separate sheets of the map are not computed independently by general tables, but all are computed with respect to the one chosen prime meridian and fixed centre C upon it. Hence the sheets of the map fit accurately together.

The projection has no scientific interest, but it is very useful and suitable for a topographical series of a comparatively small country. [V*]

Its principal defects are (1) that the scale north and south is exaggerated on each side of the central meridian. It should not be used for maps extending more than 200 miles from the central meridian; (2) that the meridians away from the centre are not straight; and (3) that as used in the British Ordnance Survey, the sheets are made rectangular; they are not bounded by meridians and parallels; and the edges of the sheets are not in general north and south, but may be inclined as much as 4° to the meridians. This is a very fruitful cause of mistakes.

We may take it as a general principle that survey sheets should always be bounded by meridians and parallels. [W*]

Mollweide's homolographic projection.

This projection has been used very frequently of late for small maps of the world. It is equal area, and therefore suitable for distribution maps. It is neat in shape—an ellipse having the major axis twice the minor; but the distortion near the extremities of the minor axis is necessarily extreme.

Consider first the construction of a hemisphere on this projection. Take $r = \sqrt{2} . R$ where R is the Earth's radius. Then a circle with radius r has the same area as the hemisphere.

Take a diameter of this circle to represent the half of the equator; and divide it equally into say six parts, each representing 30° of longitude. Draw ellipses through the poles and these points on the equator, to represent meridians. By an elementary property of the ellipse, the areas of these "gores" are all equal.

The parallels of latitude are straight lines parallel to the equator. The distance from the equator of a parallel of latitude ϕ is $r \sin \theta$, where

$$\pi \sin \phi = 2\theta + \sin 2\theta.$$

This equation cannot be solved directly, to find θ where ϕ is given; but by the reverse process it is easy to find ϕ for any given value of θ; and when this has been done for a sufficient number of cases, the values of θ corresponding to any desired value of ϕ can be interpolated. In this way the table on p. 121 was constructed.

Having drawn the circle, and the parallels of latitude by means of the table, we divide all the parallels into the same number of equal parts as the equator; and curves drawn through the series of corresponding points give us the elliptical meridians. Hence the projection is quite easy to construct. And by extending the parallels on each side, with the same equal divisions, we obtain the representation of as much as is desired of the other hemisphere.

It is easy to show that the projection is equal area. For on our projection the area of a belt of the hemisphere between the equator and latitude ϕ is easily seen to be

$$\tfrac{1}{2}r^2 \sin 2\theta + r^2 . \theta$$
$$= R^2 (\sin 2\theta + 2\theta) = \pi R^2 \sin \phi,$$

which is the area of the corresponding belt of the hemisphere. Hence the belts between parallels are of their true areas. And we have already seen that the gores between meridians are also true in area. Hence the projection is equal area.

On this projection the world is generally represented as an ellipse with the equator as major axis. But the frontispiece of the Catalogue of Maps published by the Topographical Section, General Staff (July 1908) is a beautiful example of a transverse Mollweide*, with major axis the meridian 70° E.—110° W. The greater part of the British Empire falls within the region of not too great distortion, and the relative areas of its parts are excellently shown.

---

* Reproduced as the frontispiece of this book by permission of Colonel Close, R.E.

Aitoff's projection of the whole sphere.

This is a somewhat new and interesting projection which resembles Mollweide's, but has some points of superiority. The method of construction is as follows :

Take the zenithal equal area projection of a hemisphere, with zenith on the equator ; and pass through the straight line representing the equator a plane making an angle 60° with the plane of the projection. Project the net of the zenithal projection orthographically on this plane. We then have a projection of the hemisphere bounded by an ellipse of which the major axis is twice the minor. Now halve the scale in longitude: that is to say, number the meridian which was 10° from the central meridian as 20° ; and so on. We thus obtain a representation of the whole sphere within the boundaries of the ellipse.

The projection is evidently equal area, since we started with an equal area projection, and this property is not modified by the orthogonal projection on to the plane. And it has the advantage over Mollweide's that the angles of intersection of the meridians and parallels are not so greatly altered towards the eastern and western edges of the sheet.

It can be constructed very readily when we have a table of the rectangular coordinates of the intersections for the zenithal projection, by plotting the x's unchanged and halving the y's.

It is clear that there is a whole class of projections of this kind that might be constructed. But it is only the equal area property that is preserved unaltered in the orthogonal projection on the plane, and only the 2 : 1 reduction that offers any special convenience.

Breusing's projection.

This is an attempt to obtain a mean between the advantages of the zenithal equal area and the zenithal orthomorphic projections. The radii are the geometrical means between the radii of those two projections, or

$$r = 2 \sqrt{\tan \tfrac{1}{2}\zeta \sin \tfrac{1}{2}\zeta}. \quad [X^*]$$

This formula gives distances from the centre slightly greater than the true distances, but not so exaggerated as in the

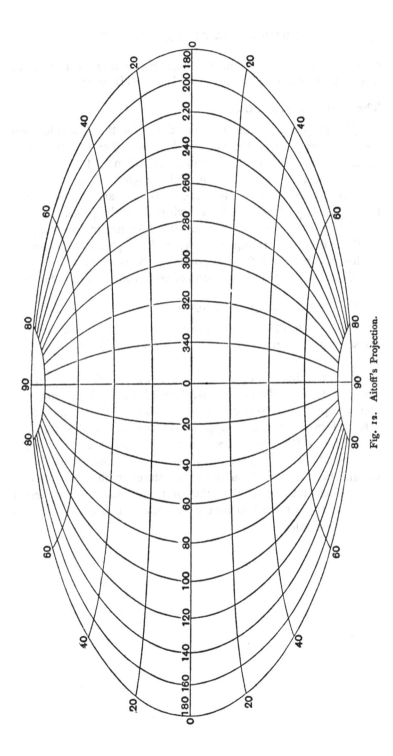

Fig. 12. Aitoff's Projection.

orthomorphic. It is not much used, and is of no special interest. The radii for each 10° are given in Table VIII, p. 120.

The globular projection.

This projection, often used in atlases for the World in Two Hemispheres, is very simple, but has no other merits. It is constructed as follows : The central meridian and the equator are two equal straight lines at right angles. The equator is divided into equal parts ; and the meridians are arcs of circles passing through these points of division, and the poles.

The central meridian, and the circumference of the map, are similarly divided into equal parts ; and the parallels are arcs of circles passing through the points of division of the central meridian, and corresponding pairs of points on the circumference.

Nell's modified globular projection.

This is a kind of mean between the ordinary globular and the stereographic projection of the hemisphere. It is constructed as follows :

The meridians and parallels are curves drawn midway between the meridians and parallels of the globular and the stereographic projection of the hemisphere. Thus the meridians intersect in the poles, at opposite extremities of the diameter which forms the central meridian. The parallels of latitude intersect the bounding circle in those points in which it is divided equally both by the globular and the stereographic projection.

It does not appear that the projection has any particular merit, and it is described here only because it is occasionally to be found in atlases.

CHAPTER VII

PROJECTIONS IN ACTUAL USE

WE have already remarked that the projections which are the most often described in the elementary accounts of the subject are in general those which are the least frequently met with in actual use. In order to get some idea of the relative usefulness, or at any rate, of the relative use of the various projections that we have studied, we will collect some statistics of the projections employed on the sheets of the principal topographical surveys of the world, and of the principal atlases in which attention has been paid to the choice of the projection.

Projections employed on topographical maps.

It is quite impossible to discover by measurement on the sheet what is the projection employed, since the irregular expansion of the paper much more than conceals the minute differences that there are between one projection and another in the small area covered by a single sheet of a topographical map.

I am indebted to Colonel C. E. Close, C.M.G., R.E., Director-General of the Ordnance Survey, for the information which forms the basis of the following list, and to Colonel Hedley, R.E., Chief of the Geographical Section of the General Staff, for some additions to it.

Austria-Hungary ...　1/75,000.　Polyhedric.
　　　　　　　　　　　1/750,000.　Old edition.　Bonne.
　　　　　　　　　　　　　　　　New edition.　Albers' conical
　　　　　　　　　　　　　　　　equal area.

Belgium ... ... 1/20,000 and reductions. Bonne.
Denmark ... ... 1/20,000. Old map. Bonne.
1/40,000. Jutland. A modified conical.
All recent Staff maps simple conic with
standard parallel 56° N.
Egypt ... ... "Gauss' conformal." See Appendix.
France ... ... 1/80,000 and 1/200,000. Bonne.
1/100,000. Polyhedric.
1/50,000. New map. Polyconic.
Germany ... ... 1/25,000 and reductions. Polyhedric.
Italy ... ... 1/100,000. * "Polycentric or natural"
(? polyhedric).
1/500,000. Bonne.
Netherlands ... 1/25,000. Bonne.
Norway ... ... 1/100,000. * "Crescent or increasing
conic" (? conic with 2 st. parallels).
Russia ... ... 1/126,000. Bonne.
1/420,000. Conical orthomorphic.
Spain ... ... 1/50,000. Probably polyhedric.
1/200,000. Bonne.
Sweden ... ... 1/100,000. See Norway.
Switzerland ... 1/25,000 and 1/50,000. Bonne.
United Kingdom—Ordnance Survey
England 1/2,500, 1/10,560, 1/63,360, 1/126,720,
1/253,440. Cassini.
1/633,600. Airy's projection by Balance
of Errors.
Scotland and Ireland 1/63,360 and smaller. Bonne.
United Kingdom—Geographical Section of the General Staff.
Early maps : Rectangular polyconic.
Recent maps : Polyconic and Inter-
national.
United States ... 1/62,500. Polyconic.

We notice that the projection of Bonne, or of the Dépôt
de la Guerre, is very much used ; but this must be considered
rather as testimony to the great reputation of that department
than as evidence of the suitability of the projection for the

---

* Official description : identification uncertain.

sheets of a topographical series. A single sheet on this projection, plotted on its central meridian, is almost if not quite indistinguishable from the corresponding sheet of the polyconic; while if a number of sheets are plotted as one whole the inclination of the meridians to the parallels is a grave defect.

The Service Géographique de l'Armée, which has succeeded to the survey duties of the Dépôt de la Guerre, has recently abandoned Bonne's projection for the polyconic and its elegant modification the projection of the International Map. It seems probable that these will be generally employed in new work to the exclusion of all others.

Projections employed in Atlases.

Until a few years ago it was rare to find on the margin of an Atlas map any statement of the projection on which it was constructed. Of recent years much more interest has been taken in the choice of the projection, and it is usual to find such a statement in the best French and German Atlases. We must regret that it is still uncommon in the Atlases of British publishers.

We will examine the following atlases, and see what projections are used in them for the principal continental maps:

1. E. Debes. Neuer Handatlas. Leipzig, 1897.
2. C. Diercke. Schul-Atlas für höhere Lehranstalten. Braunschweig, Westermann, 1908.
3. G. Philip. Systematic Atlas. Edited by Ravenstein. London, 1894.
4. E. Schrader. Atlas de Géographie Moderne. Hachette, Paris, 1896.
5. E. Schrader. Atlas de Géographie Historique. Hachette, Paris, 1896.
6. E. Stanford. London Atlas of Universal Geography. London, 1904.
7. Stieler's Hand-Atlas. Gotha, Justus Perthes, 1905.
8. Sydow-Wagner's Methodischer Schul-Atlas. Gotha, Justus Perthes, 1908.
9. Vidal-Lablache. Paris, Armand Colin, 1894.

10. St Martin-Schrader. Atlas Universel de Géographie. Paris, Hachette, 1877–1912.

In numbers 1, 2, 3, 4, 5, and 8 the name of the projection is given on the margin or in the introduction; in numbers 6, 7, 9, and 10 this is not done, and the identifications given below are subject to a little uncertainty in some cases.

In these atlases the following projections are used:

EUROPE.	Debes	Conical orthomorphic
	Diercke	Zenithal equal area
	Philip	Bonne
	Schrader Mod.	Bonne
	Schrader Hist.	Simple conic
	Stanford	Simple conic
	Stieler	Bonne
	Sydow-Wagner	Bonne
	V.-Lablache	Bonne
	St Martin	Bonne
ASIA.	Debes	Zenithal equidistant
	Diercke	Zenithal equal area
	Philip	Zenithal equal area
	Schrader Mod.	Zenithal equidistant
	Schrader Hist.	Werner
	Stanford	Bonne
	Stieler	Bonne
	Sydow-Wagner	Bonne
	V.-Lablache	Bonne
	St Martin	Zenithal equidistant
AFRICA.	Debes	Zenithal equidistant
	Diercke	Zenithal equal area
	Philip	Sanson-Flamsteed
	Schrader Mod.	Sanson-Flamsteed
	Stanford	Sanson-Flamsteed
	Stieler	Zenithal equidistant
	Sydow-Wagner	Sanson-Flamsteed
	V.-Lablache	Sanson-Flamsteed
	St Martin	Zenithal equidistant

NORTH AMERICA.	Debes	Breusing's Zenithal
	Diercke	Zenithal equal area
	Philip	Zenithal equal area
	Schrader Mod.	Zenithal equidistant
	Stanford	Simple conic
	Stieler	Bonne
	Sydow-Wagner	Bonne
	V.-Lablache	Bonne
	St Martin	Zenithal equidistant
SOUTH AMERICA.	Debes	Breusing's Zenithal
	Diercke	Bonne
	Philip	Sanson-Flamsteed
	Schrader Mod.	Sanson-Flamsteed
	Stanford	Sanson-Flamsteed
	Stieler	Sanson-Flamsteed
	Sydow-Wagner	Bonne
	V.-Lablache	Bonne
	St Martin	Zenithal equidistant
AUSTRALIA.	Debes	Conical orthomorphic
	Philip	Sanson-Flamsteed
	Schrader Mod.	Bonne
	V.-Lablache	Bonne
	St Martin	Bonne
OCEANIA.	Diercke	Zenithal equal area
	Schrader Mod.	Mercator
	Stieler	Zenithal equidistant
	V.-Lablache	Sanson-Flamsteed
	Sydow-Wagner	Sanson-Flamsteed
	St Martin	Mercator

POLAR REGIONS. All use polar zenithal equidistant or zenithal equal area except St Martin, who uses what seems to be a slightly oblique zenithal equidistant.

THE WORLD. Mercator, Globular, and Mollweide are used by nearly all the atlases. Debes has its hemisphere maps on the zenithal equal area, instead of the familiar Globular. Philip's Systematic Atlas has small examples of James' and Airy's projections of more than the hemisphere;

and Schrader's several atlases make considerable use of Aitoff's projection, which is uncommon, and of the Orthographic.

It will be seen from the above lists that the number of projections in common use is quite small, and that many of those which are móst frequently described:—conical equal area, stereographic, or orthographic are scarcely found in use at all. The author has not come across an example of the simple conical equal area, but it is stated by Bludau that there is an example of it in Lüddecke's Deutsche Schul-Atlas in the United States map. Albers' conical equal area projection with two standard parallels does not seem to have been used in any atlas, but is adopted in the Austrian General Staff map of Central Europe, and in a wall map of Russia published by the Russian Geographical Society in Russian.

The conic with two standard parallels is found in Debes' map of Central Europe; two examples of the equatorial cylindrical orthomorphic in the same, Russia and Central America; and an oblique cylindrical orthomorphic, S.E. Asia, and an oblique zenithal orthomorphic or stereographic, for Equatorial Africa, are also found in Debes. [Y*]

The identification of projections.

From the foregoing chapters it is evident that we can construct a key which will provide for the identification of the more common projections, whenever they are shown of sufficient extent. It is usually possible to identify the projections of Atlas maps; and generally difficult to say with any certainty what is the projection on which a topographical sheet is drawn, because in the latter case the precise measurement required is forbidden by the stretch of the paper.

The following rules will serve as a guide:

1. If the parallels are concentric circles, and
    (*a*) the meridians are curved, the projection is Bonne's;
    (*b*) the meridians are straight lines, it is one of the conical.

    (A) parallels equidistant—Simple conic; or

    (B) conic with two standard parallels; not easy to distinguish from (A), but very uncommon;

    (C) distance between parallels decreasing towards the pole, and increasing away from it—Conical equal area;

    (D) distance between the parallels increasing towards the pole, and decreasing away from it—Conical orthomorphic.

2. If the parallels are straight lines, and

   (*c*) the meridians are curved, and

    (E) the parallels are equidistant—Sanson-Flamsteed;

    (F) the parallels are closer towards the poles—Mollweide.

   (*d*) the meridians are straight lines, and

    (G) the parallels are equidistant—Simple cylindrical, or "plate-carré";

    (H) the parallels are closer towards the poles—Cylindrical equal area;

    (K) the parallels are wider apart towards the poles—Mercator.

3. If the meridians are straight lines, and the parallels are curved—Gnomonic. [Z*]

4. If both meridians and parallels are curved, and

    (L) they cut one another at right angles—Zenithal orthomorphic, probably the Stereographic; or perhaps, on a topographical sheet, the Rectangular polyconic;

    (M) they do not cut one another at right angles, then it is not always possible to distinguish at sight between the zenithal equidistant, zenithal equal area, and some of the more uncommon projections such as Airy's or Clarke's. If it is an atlas map of a continent, it is most likely a zenithal, and by measuring the distances between parallels along the straight central meridian one may usually distinguish between the zenithal equidistant—central meridian divided truly; the zenithal equal area—

divisions decreasing away from the centre; and the zenithal orthomorphic—divisions increasing away from the centre. If it is the map of a hemisphere or more, then it may possibly be Airy's or Clarke's. To identify these with certainty it is necessary to make measures and compare with the formulae for the most likely cases. This is interesting but tedious, and shrinking of the paper makes the result uncertain.

It will be understood that this key is not infallible, because it takes no account of the possibility of meeting occasionally such unusual projections as the oblique cylindrical orthomorphic. It will, however, serve in the majority of cases to identify the projections that are in use.

The choice of a projection.

It is difficult to lay down rules for the choice of projections for different purposes, since the conditions which must govern the choice are so exceedingly varied. We can do no more than indicate some of the points which may be kept in view.

Within the limits of a single topographical sheet there is little difference between the merits of several projections. The choice is guided by considerations of convenience and economy of time in the drawing office. Each sheet should be bounded by meridians and parallels and should fit its neighbours along the edges. We can hardly do better than choose the polyconic, or its elegant modification, the projection of the International Map. [a*]

In choosing the projection for an atlas map, we shall do well to remember that the conical projections have these incontestable advantages: (1) the meridians are all straight lines, and the parallels concentric circles, so that the properties of the projection are the same all along the parallel; it does not deteriorate as one gets away from the central meridian; and at any point it is easy to measure with the protractor the apparent bearing of any ray. These advantages are not shared by any of the projections which have curved meridians.

The conical projection with two standard parallels affords

wide opportunities for selecting the standard parallels to suit the map; and when the whole of the sheet is north or south of the equator, it will usually be found that this, or the equal area projection with two standard parallels (Albers'), is hard to beat.

If, however, the sheet crosses the equator the conical projection becomes less suitable; and then we may find it better to adopt the zenithal equal area. Or, if the equator nearly bisects the sheet, then Sanson-Flamsteed gives good results, and is very easy to draw.

For maps of a hemisphere the choice is wide. Mollweide (for equal areas), Airy's projection by balance of errors, and Clarke's perspective projection with the appropriate distance of the centre of projection, are all excellent. These all represent shapes pretty well; and a good representation of bearings, except from the centre, can hardly be expected on a map covering so large an area.

For the whole world on a single sheet we have Mollweide, which is useful for distribution diagrams, but can scarcely be called a map; and Aitoff's, which is not so easy to draw as Mollweide's, but is better in that its representation of the shape of countries far east and west of the central meridian is not so distorted, since the meridians and parallels are not so oblique to one another. The problem of showing the sphere on a single sheet is intractable. We have examined the considerable advantages of the cubical gnomonic projection; and we may notice that it would not be difficult to re-draw on the zenithal equal area projection the six trapeziums of the sphere which correspond to the six faces of the cube. These six sheets would have a rolling fit along their edges, and would make a useful map; it has not been done, to the knowledge of the writer.

Finally, for a nautical chart, and for no other purpose whatever, we should use Mercator's projection. [β*]

CHAPTER VIII

THE SIMPLE MATHEMATICS OF PROJECTIONS

In the preceding chapters we have given a generalized, somewhat superficial, descriptive, but unmathematical treatment of the subject. We must now proceed to the simple mathematical treatment of the theory of projections.

As has been already stated, this book does not profess to attack the subject from the mathematician's point of view. We shall not attempt to examine the general theory of Gauss, how far, and by what means, a representation of form upon any surface whatever may be transformed into a similar representation upon any other surface. The mathematics of this most general case is difficult, and the results have no very obvious application to map making. Neither shall we try to show how the expressions for the more complicated projections have been, in the first instance, deduced from general theory. That has been done in an excellent way by Germain. But it is more interesting to mathematicians than to map makers or map users. For the latter people, for whom this book is written, it is sufficient to show how the properties of a projection may be proved when its formula is given, and how it may be constructed either graphically or numerically.

The general theory is hard; the theory for each particular case is easy. We shall confine ourselves to the latter, and have therefore called this chapter The Simple Mathematics of Projections.

Standard notation.

In order to avoid continual repetition of definitions, we shall find it convenient to adopt a uniform notation for the various quantities which enter into the expressions.

We shall suppose that

R = the radius of the Earth, supposed spherical (or the equatorial radius, if we are taking account of obliquity), expressed in the scale units of the intended map. For example, if the map is to be on the scale of one in a million R = one millionth of the actual radius of the Earth;

ϕ = the geographical or astronomical latitude;

χ = the complement of the latitude, or the co-latitude;

λ = the longitude of a point on the Earth;

$\Delta\lambda$ = the difference of longitude between two points or the angle between the meridians passing through those points.

Conical projections.

In *normal* conical projections, that is to say, in conical projections where the meridians are represented by straight lines converging to the pole, and the parallels of latitude by concentric circles,

r = the radius of a circular parallel of latitude ϕ.

If the projection has one standard parallel, that is, one parallel which is shown its true length upon the map,

r_0 = the radius,

and ϕ_0 = the latitude of that standard parallel,

or χ_0 = its co-latitude.

If the projection has two standard parallels,

$r_1,\ r_2$ = the radii,

$\phi_1,\ \phi_2$ = the latitudes of the two standard parallels,

or $\chi_1,\ \chi_2$ = their co-latitudes,

    θ = the angle between two radii representing meridians whose difference of longitude is $\Delta\lambda$,

$n =$ the constant of the cone, and $2\pi n$ the angle at the pole of the projection which corresponds to a difference of longitude of $360°$ or 2π; so that
$$\theta = n . \Delta\lambda.$$

As oblique conical projections are of no importance, it is not necessary to define the modifications in the above notation required to deal with them.

Cylindrical projections.

The normal cylindrical projection is a special case of the normal conical projection, when the pole of the projection is removed to infinity. The parallels then become straight lines; r is infinite for all the parallels; θ and n are zero. Our formulae will be adapted to give, not r_0 and r, but $r_0 - r$, the distance between any selected parallel and the standard parallel.

We shall not require to deal with oblique cylindrical projections. [γ^*]

Zenithal projections.

In these cases the pole of the projection becomes the centre of the map; h is unity.

We shall continue to use r for the radius from the pole, but must remember that in the usual case, when the projection is oblique, the concentric circles of radius r no longer represent parallels of latitude upon the Earth, but parallels of equal distance from the centre of the map.

$\zeta =$ the angular distance of such a parallel from the centre.

It will be noticed that angles are uniformly represented by small Greek letters; distances by small italic letters.

Scale.

In all our tables of radii, and in the drawings of the various projections, we shall work to the scale of one in a hundred million, or $1 : 10^{-8}$. It was intended that the metre should represent the ten-millionth part of the distance from Pole to Equator; and the relation between the Earth and the actual metre, as defined by the length of the platinum standard preserved in the Bureau International des Poids et Mesures at Sèvres, is nearly though not exactly what was intended. For our purposes it will be amply sufficient to assume that the distance from Pole to Equator of the Earth is actually 10,000,000 metres; so that if we work on the scale of one to a hundred million this distance will be represented in our tables or on our drawings by 0·1 metre or 100 millimetres.

The corresponding value of R is $200/\pi = 63\cdot66$ mm.; and an arc of $1° = R\pi/180 = 1\cdot111$ mm.

The ellipticity of the Earth.

In our elementary computations and small scale drawings it will be unnecessary to take account of the ellipticity of the Earth. On the small scale of $1 : 10^{-8}$ it is inappreciable. In any Atlas maps it can usually be neglected. In the larger Survey maps, which are usually upon some form of conical or polyconic projection, the ellipticity of the Earth is taken into account by the use of extensive geodetic tables such as those of the Indian Survey. We shall indicate the cases where these should be employed. In other cases dealt with in this book we shall assume that the ellipticity may be neglected.

Conical projections.

In all conical projections the meridians are represented by straight lines radiating from a point, the pole of the projection, and the parallels by concentric circles described about the pole. Either one or two of these parallels are made of their true length, that is to say, are true to scale, or are equal in length to the length of the corresponding parallels on the Earth multiplied by the representative fraction of the map.

Having given the radius and the length of the arc of parallel we obtain at once the angle which it subtends at the pole of the projection. If this angle is $n . 2\pi$, then n is what we have called the "constant of the cone."

We shall find it convenient to examine projections systematically in the following order of properties :

(*a*) radius and length of the standard parallel, or parallels.
(*b*) constant of the cone *n*.
(*c*) radii of the other parallels.

(*d*)   linear scale along the meridians and parallels.
(*e*)   scale of areas.
(*f*)   alteration of angles.
(*g*)   construction by rectangular coordinates.

Simple conical projection with one standard parallel.

The radius of the standard parallel is the length of the tangent drawn from a point in the polar axis produced to touch the sphere at the standard parallel; that is

$$r_0 = R \cot \phi_0 \text{ or } = R \tan \chi_0 \quad \ldots\ldots\ldots\ldots(1).$$

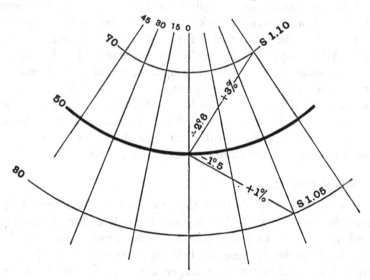

Fig. 13.   Simple Conic.

The length of an element of the standard parallel is

$$R \cos \phi_0 . \Delta\lambda.$$

Equating this to the alternative expression for its length, viz. $r_0 . \theta$ we have

$$R \cos \phi_0 . \Delta\lambda = R \cot \phi_0 . \theta,$$

whence we have for the constant of the cone

$$n = \frac{\theta}{\Delta\lambda} = \sin \phi_0 \ldots\ldots\ldots\ldots\ldots\ldots(2).$$

The lengths along the meridians are true. Hence the general expression for the radius is

$$r = r_0 - R(\phi - \phi_0)$$
$$= R\{\cot\phi_0 - (\phi - \phi_0)\} \quad\ldots\ldots\ldots\ldots\ldots(3).$$

In practice we draw the standard parallel, and lay off along the central meridian the true distances to the other parallels which can then be described. We divide the standard parallel truly; and join the points of division to the pole of the projection to obtain the other meridians. We can easily allow for the ellipticity of the Earth by taking the true distances from geodetic tables.

The only difficulty in constructing the projection graphically lies in the awkwardness of describing circles of very large radius. For this reason it is often preferable to calculate rectangular coordinates of the intersections and plot them. See p. 80. By construction the meridians are their true lengths. Hence the scale along them is true. The expression for the scale along the meridian, obtained by differentiating (3), is

$$\frac{dr}{R\,d\phi} = -1;$$

the sign is negative because r increases as ϕ decreases.

The scale along any parallel of latitude ϕ is

$$\frac{r\,d\theta}{R\cos\phi\,d\lambda} = \frac{r\sin\phi_0}{R\cos\phi}$$
$$= \{\cos\phi_0 - (\phi - \phi_0)\sin\phi_0\}\sec\phi \quad\ldots\ldots(4).$$

Since the scale along the meridians is unity, the scale of areas is the same as the scale along the parallels. [δ^*]

Computation of an example.

Suppose that we wish to make a simple conical projection for a small map of Europe, on the scale of 1 : 100,000,000.

Take the parallel of 50° N. as the standard parallel.

Its radius is $R\cot 50° = 53\cdot4$ mm. (when $R = 63\cdot66$).

The constant of the cone $n = \sin 50° = 0\cdot766$.

The radius of the parallel of 70° is

$$53\cdot4 - 20 \times 1\cdot111 = 31\cdot2 \text{ mm.}$$

and of the parallel $35°$ is $53\cdot4 + 15 \times 1\cdot111 = 70\cdot1$ mm.

The scale along the parallel 70° is $\dfrac{nr}{R\cos\phi} = 1\cdot097$, and along the parallel 35° is $1\cdot030$;
errors of 10 and 3 per cent. respectively.

Take the meridian 20° E. of Greenwich as the central meridian, and let us compute the rectangular coordinates of the point 70° N. 65° E., in the Kara Sea, just north of the boundary between Europe and Asia,

$$x = r\sin\theta = r\sin(n\cdot45°)$$
$$= 31\cdot2\sin34°\ 28'$$
$$= 17\cdot7\text{ mm.,}$$
$$y = r_0 - r\cos\theta = 53\cdot4 - 25\cdot7 = 27\cdot7\text{ mm.,}$$

and similarly any other point is computed.

The distance of this point from the centre of the map is $\sqrt{x^2+y^2} = 32\cdot9$ mm.

And its bearing from the central meridian is $\tan^{-1}\dfrac{x}{y} = 32°\ 35'$.

But the true distance and bearing are $32\cdot1$ mm.; 30° 0′. Hence our projection gives us an error of about three per cent. in distance and $2\frac{1}{2}$° in azimuth for the line from the centre to near the top right-hand corner of the map.

We shall compute other projections for this same case of a map of Europe, and be able to compare their relative merits.

A concise idea of the variation of scale along the parallel is given by the following small table for the three cases of standard parallels $22\frac{1}{2}$°, 45°, $67\frac{1}{2}$°.

Scale along the parallel. Simple conic.

Parallel ϕ	$\phi_0 = 22\frac{1}{2}°$	$\phi_0 = 45°$	$\phi_0 = 67\frac{1}{2}°$
0°	1·074	—	—
10	1·023	1·156	—
20	1·001	1·081	—
30	1·009	1·030	—
40	1·054	1·004	1·080
50	1·151	1·004	1·034
60	—	1·044	1·007
70	—	1·165	1·001
80	—	—	1·043

Simple conical projection with two standard parallels and true meridians.

Let AA', BB', be elements of the two standard parallels, of the same extent in longitude. We have to choose O, the pole of the projection, in such a way that AA', BB' shall be arcs of circles concentric at O; shall subtend the same angle at O; and shall be their true distances apart.

We have

$$\frac{OA}{OB} = \frac{AA'}{BB'} = \frac{\cos\phi_1}{\cos\phi_2}.$$

Hence

$$\frac{OA}{OA - OB} = \frac{\cos\phi_1}{\cos\phi_1 - \cos\phi_2},$$

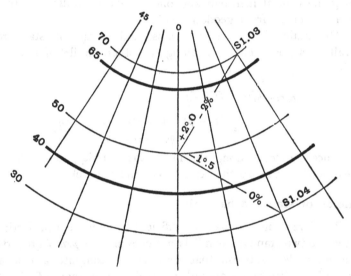

Fig. 14. Conic with two Standard Parallels.

and

$$r_1 = OA = R(\phi_2 - \phi_1)\frac{\cos\phi_1}{2\sin\frac{\phi_2 + \phi_1}{2}\sin\frac{\phi_2 - \phi_1}{2}} \quad\ldots\ldots(1),$$

which is the expression for the radius of the standard parallel.

For the constant of the cone we have

$$r_1 . \theta = R\cos\phi_1 . \Delta\lambda,$$

whence

$$n = \frac{\theta}{\Delta\lambda} = \frac{2}{\phi_2 - \phi_1}\sin\frac{\phi_2 + \phi_1}{2}\sin\frac{\phi_2 - \phi_1}{2} \quad\ldots\ldots\ldots(2).$$

The lengths along the meridians are true. Hence the general expression for the radius is

$$r = r_1 - R(\phi - \phi_1)$$

$$= R\left\{ \frac{(\phi_2 - \phi_1)\cos\phi_1}{\cos\phi_1 - \cos\phi_2} - (\phi - \phi_1) \right\}$$

$$= R\frac{(\phi_2 - \phi)\cos\phi_1 - (\phi_1 - \phi)\cos\phi_2}{\cos\phi_1 - \cos\phi_2}.$$

In practice we may draw a standard parallel by computation from (1) and then construct the projection as the simple conic with one standard parallel was constructed, by laying off along the central meridian and one standard parallel the true distances taken from geodetic tables.

The scale along the meridians, and along the standard parallels, is true. The scale along any other parallel of latitude ϕ is

$$\frac{r\,d\theta}{R\cos\phi\,d\lambda} = \frac{nr}{R\cos\phi}$$

$$= \frac{(\phi_2 - \phi)\cos\phi_1 - (\phi_1 - \phi)\cos\phi_2}{(\phi_2 - \phi_1)\cos\phi}.$$

Since the scale along the meridians is unity, the scale of areas is the same as the scale along the parallels.

Choice of standard parallels.

The foregoing formulae are sufficient to compute a projection when the two standard parallels are chosen on *a priori* grounds. For example: suppose that we wish to compute a conical projection with two standard parallels for a map of South Africa south of the Zambesi, that is to say, between the bounding parallels 15° S. and 35° S.

(*a*) We may say that we shall obtain a very fair result if we select 20° S. and 30° S. as our standard parallels; and proceed to the computation upon this assumption.

But we shall do better if, instead of selecting our parallels in this somewhat arbitrary manner, we choose them subject to more rigorous conditions. For example:

(*b*) The absolute errors along the central and the extreme

parallels between a pair of meridians may be made equal. This is Euler's projection.

(*c*) Or the errors of scale on these parallels may be made the same.

(*d*) Or the mean length of all the parallels may be made correct, while the errors on the extreme parallels, or the errors of scale on the extreme parallels, may be made equal.

(*e*) Or the maximum error of scale between the standard parallels may be made equal to the error of scale on the limiting parallels. This is not quite the same as (*b*), for the error of scale on the middle parallel is not the maximum error of scale, although near it.

In these four cases we have to find the expression for the radii and the constant of the cone without any previous knowledge of the parallels which will be of their true length, and our procedure will be somewhat different from that considered in the first place. This aspect of the problem is, however, the one which will generally occur in practice. And the conical projection with two standard parallels is such a valuable projection, so admirably suited to the smaller scale survey maps, such as the one in a million series, and yet has been until recently so comparatively little used, that we will work the same example in all the different cases defined above, for purposes of comparison.

Computation of five cases.

Computation of a conical projection with two standard parallels, with limiting parallels 15° and 35° south latitude (South Africa, south of the Zambesi).

CASE I. *Standard parallels selected arbitrarily at* 20° S. *and* 30° S.

Our general expressions give us

$$r_1 = R\,(\phi_2 - \phi_1)\,\frac{\cos\phi_1}{\cos\phi_1 - \cos\phi_2},$$

$$n = \frac{\cos\phi_1 - \cos\phi_2}{\phi_2 - \phi_1}.$$

$\phi_2 - \phi_1$ is in circular measure. R is the radius of the Earth (supposed spherical) upon the scale of our map. Taking our scale, as in all the examples, as one in a hundred million, $R = 63\cdot66$ mm. (see p. 77).

$$\text{Then} \quad n = \frac{\cos 20° - \cos 30°}{10 \times \pi/180} = \frac{0\cdot9397 - 0\cdot8660}{0\cdot1745}$$

$$= 0\cdot423,$$

$$r_1 = \frac{R \cos \phi_1}{0\cdot423} = 141\cdot6 \text{ mm.}$$

This is the radius of the parallel 20°. The radii of the others are immediately derived from it when we remember that on the scale of $1 : 10^{-8}$, $1° = 1\cdot1111$ mm.

Hence the radius of the circular parallel representing the pole, which will be found useful for comparison with the following examples, is $141\cdot6 - 77\cdot8 = 63\cdot8$ mm.

CASE II. *Absolute errors along extreme and central parallels equal (and of opposite sign).*

The absolute error in the length of any parallel is evidently

$$r \cdot \theta - R \cos \phi \cdot \Delta\lambda, \quad \text{or} \quad (rn - R \cos \phi) \Delta\lambda.$$

Let s be the radius, in millimetres, of the parallel representing the pole.

Then

$$r = [s + (\text{co-latitude in degrees}) \times 1\cdot1111] \text{ mm.}$$

And our equations are

$$n (s + 75 \times 1\cdot1111) - 63\cdot66 \cos 15°$$
$$= n (s + 55 \times 1\cdot1111) - 63\cdot66 \cos 35°$$
$$= - [n (s + 65 \times 1\cdot1111) - 63\cdot66 \cos 25°],$$

which reduce to

$$2ns + 155\cdot6n - 119\cdot19 = 0,$$
$$2ns + 133\cdot3n - 109\cdot85 = 0,$$

whence

$$n = 0\cdot419,$$
$$s = 64\cdot4 \text{ mm.}$$

CASE III. *Errors of scale along extreme and central parallels equal (and of opposite sign).*

The error of scale along any parallel is $\dfrac{rn}{R \cos \phi} - 1$.

Hence we may evidently derive our equations for this case from those in the last by dividing the respective members by the appropriate values of $R \cos \phi$, or multiplying by sec ϕ. They then become

$$n \, (z + 75 \times 1 \cdot 1111) \sec 15^\circ - 63 \cdot 66$$
$$= n \, (z + 55 \times 1 \cdot 1111) \sec 35^\circ - 63 \cdot 66$$
$$= - \, [n \, (z + 65 \times 1 \cdot 1111) \sec 25^\circ - 63 \cdot 66],$$

which reduce to

$$2 \cdot 139 \, nz + 165 \cdot 96 \, n - 127 \cdot 32 = 0,$$
$$2 \cdot 324 \, nz + 154 \cdot 29 \, n - 127 \cdot 32 = 0,$$

whence

$$n = 0 \cdot 423,$$
$$z = 63 \cdot 1 \text{ mm.}$$

CASE IV. *Mean length of all the parallels correct, and errors of scale on the extreme parallels equal.*

If the mean length of all the parallels is correct (and the lengths of all the meridians are true, by hypothesis), it follows that the total area of the map is true. The condition for this is evidently

$$\tfrac{1}{2} \, (r_1{}^2 - r_2{}^2) \, . \, \theta = R^2 \, (\sin \phi_2 - \sin \phi_1) \, \Delta \lambda,$$

or $\quad (r_1 + r_2) \, (r_1 - r_2) \, n = 2R^2 \, (\sin \phi_2 - \sin \phi_1),$

which for our example becomes

$$(2z + 130 \times 1 \cdot 1111) \, (20 \times 1 \cdot 1111) \, n = 2 \times 63 \cdot 66^2 (\sin 35^\circ - \sin 15^\circ),$$

or $\qquad\qquad 44 \cdot 44 \, nz + 3210 \, n - 2552 = 0.$

The condition that the errors of scale along the extreme parallels are equal gives us, as in Case III,

$$2 \cdot 256 \, nz + 160 \cdot 84 \, n - 127 \cdot 32 = 0,$$

which is so nearly equivalent to the preceding equation that it is evident that, without keeping more significant figures, it is impossible to solve for nz and n with any exactness.

The explanation is obvious. For the map under consideration, the solution of any of the preceding cases so nearly satisfies also the condition that the mean length of all the parallels is correct, that the latter practically introduces no new factor.

CASE V. *Maximum error of scale between the standard parallels equal to the error of scale of the extreme parallels.*

The error of scale between the standard parallels is

$$1 - \frac{nr}{R \cos \phi},$$

which is most different from unity when $\dfrac{nr}{R \cos \phi}$ is a maximum.

Differentiating with respect to ϕ and remembering that

$$dr = - R d\phi,$$

we have as the condition for maximum

$$\cot \phi = \frac{r}{R} = \frac{z + (90° - \phi) \times 1\cdot1111}{R}.$$

Hence, as in Case III, if the maximum error of scale between the parallels is equated to the error of scale on the extreme parallels, we have to find n, z and ϕ from the equations

$$n(z + 75 \times 1\cdot1111)\sec 15° - 63\cdot66$$

$$= n(z + 55 \times 1\cdot1111)\sec 35° - 63\cdot66$$

$$= -[n(z + (90° - \phi) \times 1\cdot1111)\operatorname{cosec}\phi - 63\cdot66]$$

combined with the equation for $\cot \phi$ above.

These equations are awkward to solve, and we need not give the steps of the solution here. The solution has been given by Colonel Close, *Text Book of Topographical Surveying*, p. 108, with the following result (transformed into our notation):

$$n = 0\cdot424,$$

$$z = 62\cdot9 \text{ mm.},$$

$$\phi = 25° \ 20',$$

which is exceedingly close to the values we have found in Cases I to III.

A comparison of these five cases shows very clearly that, for a map between the parallels 15° and 35° the conical projection with two standard parallels is so nearly true over the whole extent of the map, that it is almost immaterial which conditions we select for precise fulfilment.

If we take $n = 0.423$, $s = 63.1$, the solution of Case III, the error of scale on the limiting parallel of $15°$ is

$$\frac{0.423 (63.1 + 75 \times 1.1111)}{63.66 \cos 15°} - 1 = 0.008,$$

or less than one per cent., which is about the ordinary error in a printed map caused by contraction or expansion of the paper. Hence the accuracy of this projection, in this case, is as great as can be obtained practically.

For maps of greater extent in latitude, or further removed from the equator, the maximum errors of scale are naturally greater. In general it will be found sufficient to take the standard parallels about one-seventh of the whole extent in latitude from the bounding parallels. When the highest degree of refinement is required, it may be worth while to solve as in Case V. For an excellent discussion of such an example, reference may be made to a pamphlet by Colonel Close: *On the Projection for the Map of the British Isles on the scale* $1/1,000,000$. 1903.

As an example in higher latitudes we will compute the projection for the map of Europe already done for the simple conic.

The extent of latitude is roughly $70°$ to $35°$. Take then $65°$ and $40°$ as the standard parallels.

$$\text{We have} \qquad \log R = 1.8039$$
$$\log 25 = 1.3979$$
$$\log \pi/180 = \bar{2}.2419$$
$$\log \cos 65° = \bar{1}.6259$$
$$\log 0.5 = \bar{1}.6990$$
$$\log \operatorname{cosec} 52\tfrac{1}{2}° = 0.1005$$
$$\log \operatorname{cosec} 12\tfrac{1}{2}° = 0.6647$$
$$\overline{\log r_1 = 1.5338}$$

$$r_1 = 34.2$$

whence the radius for parallel $50°$ is 50.9, as compared with 53.4 for the simple conic.

And the radius of parallel $70°$ is 28.6.

Also

$$\log 2 = 0\cdot3010$$
$$\log 180/25\pi = 0\cdot3602$$
$$\log \sin 52\tfrac{1}{2}° = \overline{1}\cdot8995$$
$$\log \sin 12\tfrac{1}{2}° = \overline{1}\cdot3353$$
$$\log n = \overline{1}\cdot8960$$

$$n = \cdot787 \text{ as compared with } \cdot766.$$

The scale along the parallel 70° is 1·034, compared with 1·097.

The coordinates of the point 70° N. 65° E. are

$$x = 16\cdot6, \quad y = 26\cdot6,$$

whence the distance and bearing from the centre of the map are 31·4 mm., 31° 58′; against 32·9 mm., 32° 35′ for the simple conic; and 32·1 mm., 30° 0′ the true distance and bearing: a small but sensible improvement.

It is hardly possible to give, within the scope of this book, tables which shall fully illustrate the numerical properties of a sufficient number of cases of this projection.

Simple conical equal area (Lambert's fifth).

The distances between the concentric circular parallels are no longer true, having been modified to make the projection equal area.

The radius of the standard parallel is given by

$$r_0 = 2R \tan \tfrac{1}{2}\chi_0 \quad \dots\dots\dots\dots\dots\dots(1).$$

Equating the two expressions for the length of an element of the parallel we have

$$r_0 \cdot \theta = R \sin \chi_0 \cdot \Delta\lambda,$$

whence

$$n = \frac{\theta}{\Delta\lambda} = \frac{R \sin \chi_0}{2R \tan \tfrac{1}{2}\chi_0} = \cos^2 \tfrac{1}{2}\chi_0 \dots\dots\dots\dots(2).$$

The general expression for the radius of any parallel is

$$r = 2R \sec \tfrac{1}{2}\chi_0 \sin \tfrac{1}{2}\chi \dots\dots\dots\dots\dots(3),$$

and the projection is constructed by computing the standard parallel from (1) and dividing it truly; this gives the meridians. The radii of the parallels are then computed by (3).

The scale along a meridian is

$$\frac{dr}{R\,d\chi} = \sec \tfrac{1}{2}\chi_0 \cos \tfrac{1}{2}\chi \dots\dots\dots\dots\dots(4),$$

by differentiating (3).

The scale along a parallel is

$$\frac{r\,d\theta}{R \sin \chi . d\lambda} = \frac{nr}{R \sin \chi} = \frac{\cos \tfrac{1}{2}\chi_0}{\cos \tfrac{1}{2}\chi}\dots\dots\dots\dots(5),$$

from (2) and (3).

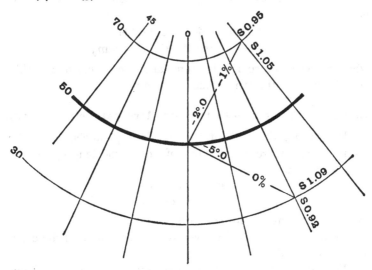

Fig. 15. Simple Conical Equal Area.

The scale of areas is

$$\frac{r\,dr\,d\theta}{R^2 \sin \chi . d\chi\,d\lambda} = \text{unity},$$

from (4) and (5).

Hence the projection is an equal area projection.

As an example we will compute the projection for the map of Europe, as we have already done for the preceding projections.

The standard parallel is 50° N.; hence $\chi_0 = 40°$; and

$$r_0 = 2R \tan \tfrac{1}{2}\chi_0 = 46\cdot3 \text{ mm.,}$$

$$n = \cos^2 \tfrac{1}{2}\chi_0 = \cdot883.$$

The radius of parallel 70° is

$$2R \sec \tfrac{1}{2}\chi_0 \sin \tfrac{1}{2}\chi = 23\text{·}5.$$

The scale along the meridian is $\sec \tfrac{1}{2}\chi_0 \cos \tfrac{1}{2}\chi$, whose value is 1·048 in lat. 70° and 0·944 in lat. 35°.

The scale along the parallel is the reciprocal of the scale along the meridian, since the projection is equal area; that is, 0·954 and 1·059 respectively.

Computing, as before, the rectangular coordinates of the point 70° N. 65° E. with respect to the centre of the map we have

$$x = r \sin 39° \ 44' = 15\text{·}0 \ \text{mm.,}$$
$$y = r_0 - r \cos 39° \ 44' = 28\text{·}2 \ \text{mm.,}$$

whence the distance is 31·9 mm. and the azimuth 28° 1', as compared with the true values 32·1 mm., 30° 0'.

Conical equal area with two standard parallels (Albers').

If r_1, r_2 are the radii of two standard parallels, that is, two parallels which are represented of their true length, we must evidently have

$$\left. \begin{aligned} r_1 &= kR \cos \phi_1 \\ r_2 &= kR \cos \phi_2 \end{aligned} \right\} \ \dots\dots\dots\dots\dots(1)$$

and the constant k is to be determined.

Whatever form may be given to k, the constant of the cone n is the reciprocal of k.

For equating, as usual, the two expressions for the length of an element of parallel, we have

$$r_1 . \theta = R \cos \phi_1 . \Delta\lambda,$$

whence
$$n = \frac{\theta}{\Delta\lambda} = \frac{1}{k} \ \dots\dots\dots\dots\dots(2).$$

It was shown by Albers that if the general expression for the radius of any parallel is

$$r^2 = 2R^2 k (\sin \phi_1 - \sin \phi) + r_1^2 \ \dots\dots\dots(3)$$

and
$$k = \frac{1}{\sin \dfrac{\phi_1 + \phi_2}{2} \cos \dfrac{\phi_1 - \phi_2}{2}} \ \dots\dots\dots(4),$$

then the projection is equal-area.

To construct the projection we must first decide upon the parallels which we shall choose as standard; then from their latitudes compute k, and thence r_1 or r_2. Draw one of the standard parallels, divide it truly, and obtain the meridians as usual. The radii of other parallels may be computed from the equation (3), or from the corresponding equation

$$r^2 = 2R^2k (\sin \phi_2 - \sin \phi) + r_2^2 \ \dots\dots\dots\dots(5).$$

But the computation is simplified if we combine (3) and (5) and obtain, after some reduction,

$$r^2 = \tfrac{1}{2} (r_1^2 + r_2^2) + 2R^2 (1 - k \sin \phi) \ \dots\dots\dots(6).$$

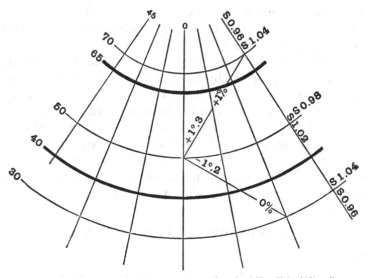

Fig. 16. Conical Equal Area with two Standard Parallels (Albers').

The scale along a meridian is

$$\frac{dr}{Rd\phi} = -\frac{kR \cos \phi}{r} \ \dots\dots\dots\dots(7),$$

which is obtained very simply when once r has been computed.
The scale along a parallel is given by

$$\frac{rd\theta}{R \cos \phi \, d\lambda} = \frac{nr}{R \cos \phi} = \frac{r}{kR \cos \phi} \ \dots\dots\dots(8).$$

And by multiplying together (7) and (8) we have for the area scale at any point

$$\frac{r\,dr\,d\theta}{R\cos\phi\,d\phi\,d\lambda} = -\,1,$$

which shows that the projection is an equal area projection.

But it will be noticed that the equal area property does not depend upon the form adopted for k. In fact we have not, up to the present, except in (6), made any use of the expression for k given in (4). We have shown that for any value of k the projection is equal area, in the sense that any two equal areas upon the Earth will be represented by two equal areas upon the map. It remains, however, to be seen whether the area scale corresponds to the linear scale upon which the standard parallels are represented.

Consider the whole area between the two standard parallels. It is $\pi n\,(r_1{}^2 - r_2{}^2)$, which by substitution for r_1 and r_2 becomes $k\pi R^2\,(\cos^2\phi_1 - \cos^2\phi_2)$. Hence the area of the map varies with k, while the lengths of the standard parallels are, it will be noticed, independent of k. It is therefore evident that there can be only one value of k which makes the area scale of the map correspond to the linear scale along the standard parallels; and this is found by equating the above area to the corresponding area of the sphere, namely $2\pi R^2\,(\sin\phi_2 - \sin\phi_1)$.

We have then

$$k\,(\cos^2\phi_1 - \cos^2\phi_2) = 2\,(\sin\phi_2 - \sin\phi_1),$$

which reduces to

$$k = \frac{2}{\sin\phi_1 + \sin\phi_2} = \frac{1}{\sin\dfrac{\phi_1 + \phi_2}{2}\cos\dfrac{\phi_1 - \phi_2}{2}}$$

as in (4).

As an example we will compute, as in previous cases, the projection for the map of Europe.

Taking 65° and 40° as the standard parallels, we have

$$k = \frac{2}{\sin 65° + \sin 40°} = 1\cdot291$$

and $n = 0\cdot775$,

$$\log k = 0\cdot1109$$
$$\log R = 1\cdot8039$$
$$\log \cos 65° = \overline{1}\cdot6259$$
$$\log r_1 = 1\cdot5407$$
$$r_1 = 34\cdot7$$
$$r_1^2 = 1206.$$

$$r^2 = 2R^2k(\sin\phi_1 - \sin\phi) + r_1^2,$$

and if $\phi = 70°$, $\qquad r = 29\cdot3$,

if $\phi = 50°$, $\qquad r = 51\cdot7$.

Hence if the point 50° N. 20° E. is the centre of the map, as before, the rectangular coordinates of the point 70° N. 65° E. are

$$x = 29\cdot3 \sin 34° 53'$$
$$= 16\cdot8,$$
$$y = 51\cdot7 - 29\cdot3 \cos 34° 53'$$
$$= 27\cdot7,$$

whence the distance is 32·4 mm. and the bearing 31° 15′.

Conical orthomorphic projections.

Consider the general properties of a conical projection defined by the equation

$$r = m(\tan \tfrac{1}{2}\chi)^n \quad \dots\dots\dots\dots\dots\dots(1),$$

where, as usual, χ is the co-latitude and n is the constant of the cone.

It is easy to show that any conical projection of this family is orthomorphic, whatever the value of m and n. For the scale along the meridian at any point is

$$\frac{dr}{R\,d\chi} = \frac{mn(\tan \tfrac{1}{2}\chi)^{n-1}\tfrac{1}{2}\sec^2\tfrac{1}{2}\chi}{R}$$

$$= \frac{nr}{R\sin\chi} \quad \dots\dots\dots\dots\dots\dots\dots\dots(2).$$

And the scale along the parallel at any point is

$$\frac{r\,d\theta}{R\sin\chi\,.\,d\lambda} = \frac{nr}{R\sin\chi} \quad \dots\dots\dots\dots\dots(3),$$

the same as the scale along the meridian.

Also, since the projection is conical, the meridians and parallels cut at right angles. This, combined with the property we have just proved, that at any point the scale along the meridians and parallels is the same, shows that the family of projections defined by (1) are all orthomorphic, whatever the values of m and n.

We have still these constants at our disposal. The constant m is evidently a simple scale constant. If we want to make the

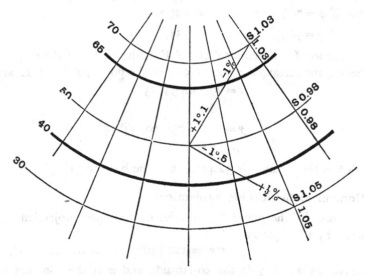

Fig. 17. Conical Orthomorphic.

parallel of co-latitude χ_0 the standard parallel, of true length, we must have

$$2\pi n \cdot m (\tan \tfrac{1}{2}\chi_0)^n = 2\pi R \sin \chi_0 \ldots\ldots\ldots\ldots(4),$$

whence
$$m = \frac{R \sin \chi_0}{n (\tan \tfrac{1}{2}\chi_0)^n} \ldots\ldots\ldots\ldots\ldots(5),$$

and our scale constant cannot be determined numerically, though we have selected our standard parallel, until we have selected further the value which we shall give to n, the constant of the cone.

But suppose, for the moment, that we have decided to give to n the same value that it has in the simple conical projection, namely $\cos \chi_0$. We shall then find the corresponding value of m from (5), and we shall have an orthomorphic projection which we may consider as constructed upon the tangent cone at the selected standard parallel.

Now the scale at any point not on this standard parallel is too large.

If then we consider any scale value, larger than the true, we shall be able to find a pair of parallels, one on each side of the standard, possessing this scale value. And it is evident that by suitably diminishing the value of the constant m we could reduce this pair of parallels to their true length; which would make the parallels between them too small in scale, while leaving the parallels outside too large. But if we start in this way, it will be a very tedious business to find, by trial and error, which pair of parallels it is that is made true by a given reduction in scale value; while if we start in the reverse way, by selecting a pair of parallels which we wish to make standard, it is evident that we cannot restrict our value of n to satisfy an independent condition as above; the two conditions will generally be irreconcilable. There is only one value of n which will make a given pair of parallels their true length, or standard. So having selected our parallels we proceed to determine the corresponding value of n, as follows.

The condition that the parallels whose co-latitudes are χ_1, χ_2 should bear their true ratio to one another is evidently, from (4),

$$\frac{(\tan \tfrac{1}{2}\chi_1)^n}{(\tan \tfrac{1}{2}\chi_2)^n} = \frac{\sin \chi_1}{\sin \chi_2}.$$

Taking logarithms of both sides, we have

$$n = \frac{\log \sin \chi_1 - \log \sin \chi_2}{\log \tan \tfrac{1}{2}\chi_1 - \log \tan \tfrac{1}{2}\chi_2} \quad \ldots\ldots\ldots\ldots(6).$$

This gives us the value of n, the constant of the cone, which makes the scale along the two selected parallels the same. To make it the true scale we must choose m so that

$$m = \frac{R \sin \chi_1}{n (\tan \tfrac{1}{2}\chi_1)^n} \quad \text{or} = \frac{R \sin \chi_2}{n (\tan \tfrac{1}{2}\chi_2)^n}.$$

It is so manifest an advantage to have two parallels their true length, instead of one, that it is usual to consider this case only of the conical orthomorphic projections. It is Lambert's Second Projection, but more generally goes by the name of Gauss.

As an example we will compute, as before, the projection for the map of Europe.

If we take the standard parallels 65° and 40° N., we have

$$n = \frac{\log \sin 25° - \log \sin 50°}{\log \tan 12° 30' - \log \tan 25°}$$

$$= 0.800.$$

$\log R = 1.8039$	$\log n = \overline{1}.9031$
$\log \sin \chi_1 = \overline{1}.6259$	$n \log \tan \tfrac{1}{2}\chi_1 = \overline{1}.4766$
1.4298	$\overline{1}.3797$
$\overline{1}.3797$	
$\log m = 2.0501$	

And $$\log r = \log m$$
$$+ n \log \tan \tfrac{1}{2}\chi.$$

Hence when $\phi = 70°$,
$$r = 28.0$$
and when $\phi = 50°$, $\qquad r = 50.0$.

If, as before, we take the centre of the map at 50° N. 20° E., the rectangular coordinates of the point 70° N. 65° E. are

$$x = 16.5 \text{ mm.}, \qquad y = 27.3 \text{ mm.},$$

whence the distance is 31.9 mm. and the bearing 31° 4′, which are nearer the true distance and bearing, 32.1 mm., 30° 0′, than in any of the previous examples.

The scale of areas at this point is $\left(\dfrac{nr}{R \sin \chi}\right)^2 = 1.058$; or the areas about this point are shown about six per cent. too great, as compared with ten per cent. for the simple conic, and three and a half per cent. for the conic with true meridians and two standard parallels. [ε*]

Zenithal projections derived from the conical.

We have already treated these projections in a descriptive way in Chapter IV. We must now consider them more formally, and will start with the zenithal equidistant projection.

In the zenithal equidistant projection the distance and azimuth of any point from the centre of the map are correctly represented. Hence the obvious method of constructing the projection, for a chosen centre, is to compute the distances and azimuths of the points of intersection of the parallels and meridians.

In the simple case when the pole is the centre no computation is needed.

In the general case, let ϕ_0, λ_0 be the latitude and longitude of the chosen centre; ϕ, λ of any other point. And let ζ, θ be the angular distance and azimuth of this latter point from the centre. Then we have at once

$$\cos \zeta = \sin \phi_0 \sin \phi + \cos \phi_0 \cos \phi \cos (\lambda - \lambda_0) \quad \ldots \ldots (1),$$

and
$$\sin \theta = \cos \phi \operatorname{cosec} \zeta \sin (\lambda - \lambda_0) \ldots \ldots \ldots \ldots \ldots (2).$$

It is easy to compute ζ directly from (1), especially if a table of natural cosines is available. Should the computer prefer to adopt the process known as "preparing (1) for logarithmic computation" he will proceed as follows.

Take two auxiliary quantities m, ω, such that

$$\left. \begin{array}{l} \sin \phi_0 = m \sin \omega \\ \cos \phi_0 \cos (\lambda - \lambda_0) = m \cos \omega \end{array} \right\} \quad \ldots \ldots \ldots \ldots (3).$$

Then
$$\cos \zeta = m \cos (\phi - \omega) \ldots \ldots \ldots \ldots \ldots (4).$$

The auxiliary angle ω is obtained from the equation

$$\tan \omega = \tan \phi_0 \sec (\lambda - \lambda_0) \ldots \ldots \ldots \ldots (5),$$

and m from either of equations (3) when ω has been found.

This gives ζ in angle. To convert it into linear measure for plotting we have

$$r = R\rho^\circ \cdot \frac{\pi}{180^\circ} \quad \ldots \ldots \ldots \ldots \ldots \ldots (6).$$

As an example of both processes we will compute, in parallel columns, the distance and azimuth of the point 70° N. 65° E.

from the point 50° N. 20° E., already required in our examples of the conical projection.

$\log \sin \phi_0$	$\bar{1}\cdot8843$		$\log \tan \phi_0$	$0\cdot0762$	
$\log \sin \phi$	$\bar{1}\cdot9730$		$\log \sec (\lambda - \lambda_0)$	$0\cdot1505$	
	$\bar{1}\cdot8573$	$0\cdot7200$	$\log \tan \omega$	$0\cdot2267$	$\omega = 59°\ 19'$
$\log \cos \phi_0$	$\bar{1}\cdot8081$		$\log \sin \phi_0$	$\bar{1}\cdot8843$	
$\log \cos \phi$	$\bar{1}\cdot5341$		$\log \operatorname{cosec} \omega$	$0\cdot0655$	
$\log \cos (\lambda - \lambda_0)$	$\bar{1}\cdot8495$		$\log m$	$\bar{1}\cdot9498$	
	$\bar{1}\cdot1917$	$0\cdot1555$	$\log \cos (\phi - \omega)$	$\bar{1}\cdot9924$	
	$\cos \zeta$	$0\cdot8755$	$\log \cos \zeta$	$\bar{1}\cdot9422$	$\zeta = 28°\ 55'$
	ζ	$28°\ 54'$			
$\log \cos \phi$	$\bar{1}\cdot5341$				
$\log \operatorname{cosec} \zeta$	$0\cdot3158$				
$\log \sin (\lambda - \lambda_0)$	$\bar{1}\cdot8495$				
$\log \sin \theta$	$\bar{1}\cdot6994$				
	θ	$30°\ 2'$			

and the computation of θ is identical with that in the other column.

There is very little to choose between the two methods. The second is perhaps somewhat more accurate when logarithm tables of a minimum number of places are used.

Finally

$$\log \zeta° = 1\cdot4612$$
$$\log R = 1\cdot8039$$
$$\log \pi/180 = \bar{2}\cdot2419$$
$$\log r = 1\cdot5070$$

$r = 32\cdot1$ mm. for the scale of $1 : 100,000,000$.

To construct zenithal projections with facility we require extensive tables computed from the above formulae for different values of ϕ_0, such as are given by Hammer (*Die geographischen wichtigsten Kartenprojektionen*, Stuttgart 1889). Abbreviated tables of this kind are given on pp. 110, 111, sufficient to enable us to compute the examples which we require.

In the conical projections the alterations of scale were symmetrical about the meridian and the parallel of latitude. Hence we examined the scale along the meridian and the scale along the parallel of latitude to obtain an estimate of the distortion at any point of the map.

In the zenithal projections, on the other hand, the alterations of scale are symmetrical about the radial from the centre and the parallel of given distance from that centre. Hence we shall examine the errors of scale along the radial and along the parallel small circle instead of along the meridian and the parallel of latitude.

It is evident that the scale along the radial is the value of the expression $\dfrac{dr}{R\,d\zeta}$; and the scale along the parallel is

$$\frac{r\,d\theta}{R\sin\zeta.\,d\theta} \quad \text{or} \quad \frac{r}{R\sin\zeta},$$

since in all zenithal projections the azimuths are true.

We will examine first the zenithal projections derived from the conical (see Chapter IV).

Zenithal equidistant projection.

$$r = R\zeta.$$

Hence the scale along the radial is unity, that is to say, distances from the centre are represented truly, as is implied in the name of the projection. And the scale along the parallel circle is $\dfrac{\zeta}{\sin\zeta}$; or if ζ is expressed in degrees, $\dfrac{\zeta\pi}{\sin\zeta.180}$. [ζ*]

The area scale is the same as the scale along the parallel.

Zenithal equal area projection.

This is a particular case of the conical equal area with one standard parallel, $\zeta_0 = 0$, whence

$$r = 2R\sin\tfrac{1}{2}\zeta.$$

The scale along the radial is

$$\frac{dr}{R\,d\zeta} = \cos\tfrac{1}{2}\zeta.$$

The scale along the parallel is

$$\frac{r}{R\sin\zeta} = \sec\tfrac{1}{2}\zeta,$$

the reciprocal of the scale along the radial, which proves the equal area property.

Zenithal orthomorphic (Stereographic) projection.

This is a particular case of the conical orthomorphic, with one standard parallel $\zeta_0 = 0$. We saw on p. 35 that its formula is $r = 2R \tan \frac{1}{2}\zeta$.

The scale along the radial is

$$\frac{dr}{R\,d\zeta} = \sec^2 \tfrac{1}{2}\zeta.$$

The scale along the parallel is

$$\frac{r}{R \sin \zeta} = \sec^2 \tfrac{1}{2}\zeta,$$

or the same as the scale along the radial, which proves that the projection is orthomorphic.

The area scale is $\sec^4 \frac{1}{2}\zeta$.

To continue the example already computed for the conical projections and the zenithal equal area, we have found that the true distance and azimuth of the point 70° N. 65° E. from the centre 50° N. 20° E. are given by

$$\zeta = 28°\ 55',\quad r = 32{\cdot}1,\quad \theta = 30°\ 0'.$$

In the zenithal equal area projection

$$r = 2R \sin \tfrac{1}{2}(28°\ 55') = 31{\cdot}8.$$

In the zenithal orthomorphic

$$r = 2R \tan \tfrac{1}{2}(28°\ 55') = 32{\cdot}8.$$

The area scales are respectively $\dfrac{\zeta\pi}{\sin \zeta \cdot 180}$, unity, and $\sec^4 \frac{1}{2}\zeta$, or 1·044, 1·000, 0·879.

Let us now bring together the cases of this example that we have computed. We have a map of Europe with centre 50° N. 20° E., and have examined how well the different projections are able to represent the distance and azimuth of the point 70° N. 65° E. from the centre; and the area about that point. The following are the results.

	Error in dist. per cent.	Error in azimuth	Error in area scale per cent.
Simple conic	2½	2½°	10
Conic with two standard parallels	2	2°	3½
Simple conical equal area	⅔	2°	0
Albers' conical equal area	1	1¼°	0
Conical orthomorphic	⅔	1°	6
Zenithal equidistant	0	0	4½
Zenithal equal area	1	0	0
Zenithal orthomorphic	2	0	12

In the next chapter we shall compute more extensive tables upon this plan.

Bonne's projection.

We have already described the modification of the simple conical projection known as Bonne's, or the Projection du Dépôt de la Guerre. The modification consists in dividing *every* parallel truly, instead of only the standard parallel. The meridians are then formed by drawing smooth curves through the points of division of the parallels.

It is easy to show that the projection is equal area. For consider a small element of area enclosed between two neighbouring parallels, and two neighbouring meridians. On the sphere

Fig. 18.

the small element of area will be a rectangle; on the projection it will be a parallelogram upon the same base and between the same parallels, and consequently equal in area to the rectangle.

The radii of the parallels are computed as for the simple conical projection. But it is clear that there is no quantity *n*, the constant of the cone, precisely analogous to the constant in the conical projection.

We may, however, look upon *n*, not as constant, but as varying with the latitude, so that the angle at the centre of the circular parallel subtended by an extent of longitude $\Delta\lambda$ is $n \cdot \Delta\lambda$; and then, by analogy with the conical projection,

$$n = \frac{R \cos \phi}{r}.$$

The projection may be constructed graphically, subject to the usual difficulty that the centre of the circular parallels is very generally off the sheet. But the computation of the rectangular coordinates is easy.

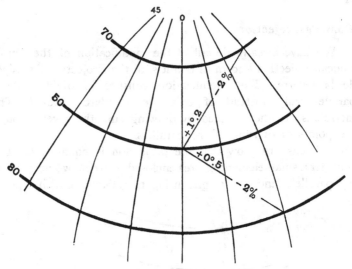

Fig. 19.   Bonne.

Let us take as an example the projection for the map of Europe already computed for the simple conic.

The standard parallel 50° N. has radius 53·4 mm.

The parallel 70° N. has radius 31·2 mm.; and for this parallel *n*, as defined above, $= R \cos \phi/r = $ 0·698.

Hence the coordinates of the point 70° N. 65° E. with respect to the centre 50° N. 20° E. are

$$x = 31 \cdot 2 \sin (0 \cdot 698 \times 45°) = 16 \cdot 3 \text{ mm.,}$$

$$y = 53 \cdot 4 - 31 \cdot 2 \cos 31° 25' = 26 \cdot 8 \text{ mm.,}$$

whence the distance from the centre is 31·4 mm. and the bearing 31° 15′, which are nearer the true distance and bearing than in the simple conic; or in the conic with two standard parallels; better than in Albers'; worse in distance but better in bearing than the simple conical equal area; and not so good as the conical orthomorphic.

These conclusions apply of course only to this particular case. They are not general. For a more general discussion of the comparative merits of these projections for a map of Europe see page 107.

Mercator's projection.

In our brief preliminary discussion of cylindrical projections (see Chap. III) we came to the conclusion that only one—Mercator's, or the cylindrical orthomorphic—required a detailed examination. $[\eta^*]$

For the general case of the conical orthomorphic with one standard parallel ϕ_0 we have

$$r = m (\tan \tfrac{1}{2}\chi)^n,$$

and

$$m = \frac{R \sin \chi_0}{n (\tan \tfrac{1}{2}\chi_0)^n}.$$

If our cone becomes a cylinder tangent to the sphere along the equator, it is clear that m becomes indefinitely large, as does also r; and that n, the constant of the cone, is zero. But in that case $mn = \dfrac{R \sin 90°}{(\tan 45°)^0} = R$, and is finite.

Remembering that $x = e^{\log_e x}$ we may write

$$r = m e^{\log_e (\tan \frac{1}{2}\chi)^n}$$

$$= m e^{n \log_e (\tan \frac{1}{2}\chi)}$$

$$= m \{1 + n \log_e (\tan \tfrac{1}{2}\chi)\} + \text{terms involving } n^2, \text{ etc.}$$

Hence $r - m = mn \log_e (\tan \tfrac{1}{2}\chi)$, since the terms in n^2 vanish; and $\qquad r_0 - m = 0.$

$$\therefore \ r_0 - r = - mn \log_e (\tan \tfrac{1}{2}\chi)$$

$$= + \frac{R}{M} \log_{10} \tan (45° + \tfrac{1}{2}\phi),$$

which is the expression we require for the distance of any parallel from the equator.

Hence we have, in rectangular coordinates,

$$x = R \frac{\pi}{180} \Delta\lambda,$$

where $\Delta\lambda$ is the difference of longitude from the central meridian expressed in degrees,

$$y = 2\text{·}30259\, R \log_{10} \tan\left(45° + \tfrac{1}{2}\phi\right).$$

The scale along a parallel is

$$\frac{dx}{R \cos\phi\, d\lambda} = \sec\phi.$$

The scale along a meridian is

$$\frac{dy}{R\, d\phi} = \tfrac{1}{2}\frac{\sec^2 \tfrac{1}{2}\chi}{\tan \tfrac{1}{2}\chi} = \sec\phi.$$

Hence the scale at any point in any direction is $\sec\phi$, and the projection is orthomorphic, which is of course obvious, because it is a special case of the conical orthomorphic projection.

As an example, let us take the map of Europe already computed for the conical projections.

When $\phi = 50°$

$$y = 2\text{·}30259 \times 63\text{·}66 \log_{10} \tan 70°$$

$$= 64\text{·}3 \text{ mm.},$$

and when $\phi = 70°$

$$y = 110\text{·}5 \text{ mm.}$$

If $\Delta\lambda = 45°$

$$x = \frac{R\pi}{4} = 49\text{·}9 \text{ mm.}$$

Hence the distance of the point 70° N. 65° E. from the point 50° N. 20° E. is 68·0 mm., or more than double the true distance; while the bearing is 47° 12′ instead of 30° 0′.

The last result shows how far from preserving the true bearings may be a projection which is formally orthomorphic.

CHAPTER IX

THE ERRORS OF PROJECTIONS

THE usual method of studying the errors of a projection is to imagine a small circle described upon the Earth, and to consider the shape which it assumes in the projection. The axes of the ellipse into which the infinitesimal circle is deformed may serve as a measure of the greatest and least changes of scale value round about the point; and a simple formula determines the maximum deformation of angle that can take place in the neighbourhood.

It seems, however, to the writer that these quantities are not really of very much value in representing the real value of the projection for the representation of areas of finite size; and no use has been made of them in this book. We propose instead to calculate some tables of the errors of distances, bearings, and areas for selected projections covering large areas, in the manner which has been illustrated already.

Let us consider first the differences between the principal members of the group of conical projections, as constructed for the map of Europe. The formulae for these projections are very various in appearance; it is instructive to see how comparatively little they differ numerically.

Our first table is constructed for the graticules of small maps on the scale of one in a hundred million. We calculate the values of n, the constant of the cone, and the radii for the three parallels of latitude 30°, 50°, and 70°. And we notice that while their absolute radii and the inclinations of the meridians vary considerably, the distances between the parallels remain very nearly the same.

Conical Projections. Radii of Parallels for Scale 1/10⁸.

On this Scale Rad. of Earth = 63·66 mm. 10° = 11·11 mm.

Projection	Standard Parallel	n	30°	Radii of Parallels Diff.	50°	Diff.	70°
Simple Conic ...	50°	0·766	75·6	22·2	53·4	22·2	31·2
Conic 2 St. Par. ...	{65°/40°}	0·787	73·1	22·2	50·9	22·2	28·7
Con. Equal Area ...	50°	0·883	67·8	21·5	46·3	22·8	23·5
Albers'	{65°/40°}	0·775	73·9	22·2	51·7	22·4	29·3
Con. Orthomorphic ...	{65°/40°}	0·800	72·3	22·3	50·0	22·0	28·0

Next, we will compute tables of the errors in distance and bearing from the centre, and of the area representation at the point, for three continental maps, of Europe, Asia, and Africa respectively.

Errors of Projections. Map of Europe. Centre 50° N. 20° E.

Errors of	Corner 70° N. 65° E. Dist. from Centre %	Az. from Centre °	Area %	Corner 30° N. 50° E. Dist. from Centre %	Az. from Centre °	Area %
Simple Conic ...	+2	+2·6	+10	+1	+8·5	+ 5
Con. 2 St. Par.* ...	−2	+2·0	+ 3	0	− 1·5	+ 4
Con. Eq. Area ...	− 1	− 2·0	0	0	− 5·0	0
Albers'* ...	+1	+1·3	0	0	− 1·2	0
Con. Orthomorph. ...	− 1	+1·1	+ 6	+1	− 1·5	+10
Zen. Equidist. ...	0	0·0	+ 4	0	0·0	+ 5
Zen. Eq. Area ...	− 1	0·0	0	− 1	0·0	0
Zen. Orthomorph. ...	+2	0·0	− 10	+2	0·0	+15
Bonne	−2	+1·2	0	−2	+0·5	0

* Standard parallels. 65° N. and 40° N.

Errors of Projections. Map of Asia. Centre 40° N. 90° E.

Errors of	Corner 70° N. 190° E. Dist. from Centre %	Az. from Centre °	Area %	Corner 20° S. 150° E. Dist. from Centre %	Az. from Centre °	Area %
Simple Conic ...	+11	+ 9·8	+26	+ 5	− 12·9	+ 53
Con. 2 St. Par.*	+ 3	+12·4	+14	− 3	− 3·7	+ 27
Con. Eq. Area ...	0	− 0·6	0	− 3	−24·1	0
Albers' * ...	+ 6	+13·2	0	− 1	− 0·6	0
Con. Orthomorph.	0	+10·2	+24	− 3	− 7·2	+ 76
Zen. Equidist. ...	0	0·0	+18	0	0·0	+ 45
Zen. Eq. Area ...	− 4	0·0	0	− 8	0·0	0
Zen. Orthomorph.	+11	0·0	+65	− 8	0·0	+208
Bonne	− 5	+10·0	0	− 12	+ 6·5	0

* Standard parallels. 63° N. and 2° N.

Errors of Projections. Map of Africa. Centre 0°, 20° *E.*

	Errors of Dist. from Centre %	Az. from Centre °	Area %
Simple Cylindrical	+ 5	+8·6	+22
Cyl. Equal Area	+ 3	+9·7	0
Zen. Equidistant	0	0·0	+22
Zen. Equal Area	− 5	0·0	0
Zen. Orthomorphic	+11	0·0	+85
Sanson-Flamsteed	− 8	+2·7	0
Clarke's Perspective	+ 2	0·0	+32

Corner 35° N. 75° E.

These are instructive as showing how rapidly the errors increase as one passes a radius of about 40°, and also how great a sacrifice of other desirable properties is entailed by the adoption of orthomorphic projections.

It will be understood that these tables refer to distances and bearings measured from the centre in each case ; and that while for the conical projections the same figures are true all along the parallel, this is not so for the zenithal projections. We must be careful, therefore, not to overvalue the zenithal projections because they make such a favourable showing when we consider them in relation to their centres.

The method of computation of these tables is as follows : The rectangular coordinates of the corners, referred to axes through the centre, are calculated, and from these the bearings and azimuths are deduced. With the conical projections it is easy to calculate the true bearing of any ray from any point, referred to the meridian through that point—easy because the meridians are straight lines. On other projections the meridians are not always straight, and it is troublesome to calculate the direction of the meridian at a point.

The distances and azimuths calculated for the projection under consideration are compared with the true distance and azimuth calculated from the spherical triangle, or taken from the tables of distance and azimuth, of which an example is given among the specimen tables at the end of the book. References are given there to more extensive collections of tables.

In comparing two projections it is often useful to place a tracing of one over the other. Thus if one has a tracing of the zenithal equidistant projection with centre at a given latitude, it is easy to find graphically the errors in distance and azimuth of any other projection for centres of that latitude.

In Figs. 13—17 and 19 we have given diagrams of the five conical projections and of Bonne, for the map of Europe; together with the numerical errors as shown in the above table.

CHAPTER X

TABLES

WE cannot attempt to give here an extended set of tables such as are required in the drawing office of a Survey Department or a Cartographer. It must suffice to give a few tables which will be useful to anyone who wishes to work out the properties of a projection.

Table I gives distances in degrees, and azimuths from centres of each 10° of latitude from the equator to 50°. These are computed by formulae such as those on page 97, and may be extended as desired in the same way. They are useful in calculating zenithal projections. The azimuths of some of the principal intersections of meridian and parallels are given at once in the table, and are the same for all the zenithal projections. The angular distance is taken from the table, and substituted in the formula which gives the corresponding radial distance for the projection in question.

In testing the accuracy of conical projections it is best to proceed as we have done in the examples: to calculate the rectangular coordinates of the intersections; thence calculate the corresponding azimuth and distance; and compare with the true azimuth and distance as given in these tables.

Should it be desired to plot a zenithal projection in rectangular coordinates, these may be computed very readily from the azimuths A and the radial distances R, since

$$x = R \sin A \text{ and } y = R \cos A.$$

TABLE I.

Distances and Azimuths from the Centre, of the Intersections of Meridians and Parallels.

Centre: Latitude 0°.

	Distances				Azimuths			
ΔL	15°	30°	45°	60°	15°	30°	45°	60°
Lat.								
10°	17° 58′	31° 29′	45° 52′	60° 30′	55° 44′	70° 34′	76° 0′	78° 30′
20	24 49	35 32	48 22	61 59	35 25	53 57	62 46	67 12
30	33 14	41 25	52 14	64 20	24 9	40 54	50 46	56 19
40	42 16	48 26	57 .12	67 29	17 9	30 47	40 7	45 54
50	51 37	56 10	62 58	71 15	12 15	22 46	30 41	36 0
60	61 7	64 20	69 18	75 31	8 30	16 6	22 12	26 34
70	70 43	72 46	76 0	80 9	5 23	10 19	14 26	17 30

Centre: Latitude +20°.

	Distances				Azimuths			
ΔL	15°	30°	45°	60°	15°	30°	45°	60°
Lat.								
−20°	42° 36′	49° 37′	59° 30′	71° 4′	158° 57′	141° 55′	129° 33′	120° 40′
−10	33 26	42 5	53 29	66 13	152 27	132 44	119 57	111 15
0	24 49	35 32	48 21	61 58	141 55	120 38	108 53	101 10
+10	17 35	30 35	44 27	58 31	122 28	104 37	96 9	90 21
+20	14 5	28 9	42 9	56 3	87 25	84 46	81 56	78 50
+30	16 51	28 51	41 43	54 42	50 39	63 46	66 58	66 47
+40	23 46	32 31	43 12	54 34	29 28	45 26	52 17	54 30
+50	32 17	38 16	46 26	55 40	18 9	31 15	38 50	42 23
+60	41 24	45 19	51 4	57. 55	11 17	20 35	27 2	30 44
+70	50 49	53 9	56 43	61 10	6 33	12 20	16 49	19 46

Centre: Latitude + 30°.

	Distances				Azimuths			
ΔL	15°	30°	45°	60°	15°	30°	45°	60°
Lat.								
−20°	52° 3′	57° 44′	66° 9′	76° 21′	162° 2′	146° 15′	133° 24′	123° 8′
−10	42 31	49 19	58 55	70 9	157 51	139 31	125 36	114 56
0	33 14	41 25	52 14	64 20	151 49	130 54	116 34	106 6
+10	24 24	34 22	46 22	59 7	141 55	119 17	105 50	96 25
+20	16 51	28 52	41 43	54 42	122 56	103 15	93 6	85 42
+30	12 59	25 54	38 43	51 19	86 14	82 22	78 18	73 54
+40	15 48	26 22	37 46	49 13	46 43	59 34	62 11	61 10
+50	22 58	30 6	39 3	48 36	25 14	39 51	46 11	47 55
+60	31 39	36 6	42 20	49 30	14 17	25 7	31 40	34 43
+70	40 53	43 25	47 13	51 50	7 46	14 24	19 14	22 8

Centre: Latitude +40°.

ΔL	Distances				Azimuths			
	15°	30°	45°	60°	15°	30°	45°	60°
Lat.								
−10°	51° 54′	57° 12′	65° 3′	74° 36′	161° 6′	144° 8′	129° 48′	117° 47′
0	42 16	48 26	57 12	67 28	157 22	138 4	122 44	110 21
+10	32 49	40 5	49 50	60 44	151 57	130 8	114 18	102 8
+20	23 46	32 31	43 12	54 34	142 53	119 3	103 57	92 49
+30	15 48	26 22	37 46	49 13	124 36	102 53	90 59	82 3
+40	11 28	22 52	34 5	45 2	85 9	80 13	75 5	69 38
+50	14 31	23 14	32 48	42 23	41 33	54 31	57 2	55 40
+60	22 5	27 20	34 9	41 34	20 8	32 59	39 1	40 44
+70	31 0	33 48	37 53	42 41	9 53	17 54	23 11	25 54
+80	40 24	41 34	43 21	45 36	3 58	7 31	10 18	12 9

Centre: Latitude +50°.

ΔL	Distances				Azimuths			
	15°	30°	45°	60°	15°	30°	45°	60°
Lat.								
−10°	61° 25′	65° 28′	71° 39′	79° 25′	163° 8′	147° 14′	132° 48′	119° 49′
0	51 37	56 10	62 58	71 15	160 43	142 59	127 28	113 51
+10	41 53	47 3	54 30	63 17	157 33	137 44	121 12	107 18
+20	32 17	38 16	46 26	55 40	152 54	130 39	113 31	99 46
+30	22 58	30 6	39 3	48 36	144 56	120 18	103 33	90 47
+40	14 32	23 14	32 48	42 23	127 51	103 56	90 11	79 45
+50	9 37	19 9	28 29	37 30	84 14	78 24	72 23	66 8
+60	13 8	19 39	27 2	34 30	34 43	48 0	51 2	49 51
+70	21 13	24 27	28 55	33 55	14 10	24 23	30 0	32 3
+80	30 26	31 40	33 33	35 53	5 5	9 31	12 50	14 52

Table II gives the lengths of degrees of the meridian and of the parallels at every 10° of latitude, in miles and in kilometres.

It will be seen that owing to the ellipticity of the Earth a degree of latitude at the Pole is nearly one per cent. longer than a degree at the Equator. And a degree of longitude at the Equator is about two-thirds per cent. longer than a degree of latitude.

This variation in the length of a degree and of a minute of arc is the source of the confusion which exists in the use of the sea or geographical mile as a unit.

TABLE II.

Lengths of Degrees of the Meridian and Parallel.

Lat.	1° of Meridian		1° of Parallel	
	Miles	Km.	Miles	Km.
0°	68·70	110·57	69·17	111·32
10	68·73	110·60	68·13	109·64
20	68·79	110·70	65·03	104·65
30	68·88	110·85	59·96	96·49
40	68·99	111·03	53·06	85·40
50	69·12	111·23	44·55	71·70
60	69·23	111·41	34·67	55·80
70	69·32	111·57	23·73	38·19
80	69·39	111·66	12·05	19·39
90	69·41	111·70	0·00	0·00

Note: The lengths of 1° on the meridian are for arcs extending half a degree north and south of the latitude named.

Table III gives the lengths of circular arcs in terms of the radius, or the circular measure of the arcs. It is useful in transforming from the angular distances of Table I to distances in some unit of length, for plotting the projection.

Thus, if the spherical distance of an intersection from the centre of the map is 17° 58′, = 17°·97, we have

$$10° = 0·1745,$$
$$7 = 1222,$$
$$0·9 = 0157,$$
$$0·07 = 0012,$$

whence $17°·97 = ·3136\ R$, where R is the radius of the Earth.

TABLE III.

Lengths of Circular Arcs, in terms of the Radius.

Deg.	Arc	Deg.	Arc
10°	0·1745	1°	0·0175
20	0·3491	2	·0349
30	0·5236	3	·0524
40	0·6981	4	·0698
50	0·8727	5	·0873
60	1·0472	6	·1047
70	1·2217	7	·1222
80	1·3963	8	·1396
90	1·5708	9	·1571

Table IV gives the radii of curvature of the meridian, and at right angles to the meridian, for each 10° of latitude.

In computing a small atlas projection it is usually sufficient to take the Earth as spherical, but it may not be quite precise enough to consider it a sphere with radius equal to the radius of the equator. We should rather consider that in the region for which we are constructing the projection it may be taken as a sphere whose radius is the mean of the radii in the meridian and at right angles to it. Thus, for a map whose centre is in latitude 50° we should take as the radius 6381 km., the mean of the quantities given in Table IV.

TABLE IV.

Radii of Curvature of the Meridian, and at right angles to the Meridian.

	Meridian	Perpendic.
	Km.	Km.
0°	6335	6377
10	6337	6378
20	6342	6380
30	6351	6383
40	6361	6386
50	6372	6390
60	6383	6393
70	6391	6396
80	6397	6398
90	6399	6399

Assumed figure $a = 6377 \cdot 4$ km.
$b = 6356 \cdot 1$ km.

Table V is extracted from the *Resolutions of the International Map Committee*, London, 1909. It suffices for the construction of any one of the sheets of the International Map on the scale of 1/1,000,000. The principles of this slightly modified polyconic projection have been described in Chap. VI, page 56.

Each sheet covers 4° in latitude and 6° in longitude.

The length of the central meridian is given in Table A. This is drawn down the centre of the sheet. Straight lines at right angles to it are drawn top and bottom. Along these the appropriate x coordinates from Table B are laid off, and the small y coordinates are erected as perpendiculars. Through the points thus constructed the top and bottom parallels of the sheet are drawn as circular arcs. The points are then joined in pairs to make the side meridians; and the intermediate parallels are drawn as circular arcs dividing the meridians equally.

The small extent of the Tables requisite for so large an enterprise is evidence of the practical convenience of the polyconic projection for sheets of this size.

TABLE V (A AND B).

TABLES FOR THE PROJECTION OF THE SHEETS OF THE INTERNATIONAL MAP OF THE WORLD.

On the scale of 1 : 1,000,000.

(Assumed figure: $a = 6378 \cdot 24$ km.
$b = 6356 \cdot 56$ km.)

TABLE V. A.

Corrected lengths on the Central Meridian in Millimetres.

Latitude		Natural length	Correction	Corrected length
From 0° to 4°		442·27	− 0·27	442·00
4	8	442·31	·27	442·04
8	12	442·40	·26	442·14
12	16	442·53	·25	442·28
16	20	442·69	·24	442·45
20	24	442·90	·23	442·67
24	28	443·13	·22	442·91
28	32	443·39	·20	443·19
32	36	443·68	·18	443·50
36	40	443·98	·17	443·81
40	44	444·29	·15	444·14
44	48	444·60	·13	444·47
48	52	444·92	·11	444·81
52	56	445·22	·09	445·13
56	60	445·52	− ·08	445·44

TABLE V. B.

Coordinates of the intersections of the Parallels of Latitude and Meridians, in Millimetres.

Lat.	Coords.	Longitude from Central Meridian		
		1°	2°	3°
0°	Mm. x	111·32	222·64	333·96
	y	0·00	0·00	0·00
4	x	111·05	222·10	333·16
	y	0·07	0·27	0·61
8	x	110·25	220·49	330·74
	y	0·13	0·54	1·21
12	x	108·91	217·81	326·73
	y	0·20	0·79	1·78
16	x	107·04	214·08	321·13
	y	0·26	1·03	2·32
20	x	104·65	209·31	313·98
	y	0·31	1·25	2·81
24	x	101·76	203·52	305·31
	y	0·36	1·45	3·25
28	x	98·37	196·75	295·15
	y	0·40	1·61	3·63
32	x	94·50	189·01	283·56
	y	0·44	1·75	3·93
36	x	90·17	180·36	270·59
	y	0·46	1·85	4·16
40	x	85·40	170·82	256·29
	y	0·48	1·92	4·31
44	x	80·21	160·45	240·73
	y	0·49	1·95	4·38
48	x	74·63	149·29	224·00
	y	0·48	1·94	4·36
52	x	68·69	137·40	206·16
	y	0·47	1·89	4·25
56	x	62·40	124·83	187·31
	y	0·45	1·81	4·06
60	x	55·81	111·64	167·52
	y	0·42	1·69	3·80

Table VI is a small specimen of the War Office Tables for the Projection of Graticules for squares of 1° side on the scale of 1/250,000, and for squares of one-half degree side on the scale of 1/125,000. These are now in use for all the sheets on these scales published by the Geographical Section of the General Staff.

The instructions for the use of these tables begin as follows:

For the 1/250,000 one degree square series.

Lay down a straight line for the central meridian, and mark off on it a length AB equal to the sum of the four 15′ lengths given in column m in the section for the latitudes between which the sheet is situated.

At A and B draw straight lines cAd, eBf at right angles to AB, and measure off lengths Ac, Ad, Be, Bf equal to half the distances given in column n opposite the latitudes of A and B respectively.

Check the total lengths cd, ef, and check the diagonal lengths cf, ed, by the value given in column q.

The ordinates of curvature, given in the last columns of the table, are then erected, and the curves of the parallels passed through them.

This provides for the bounding meridians and parallels. Those within are constructed by a method which will be obvious to any one who has followed the process up to this point.

It is important to notice that the diagonal q is the diagonal of the construction figure, and not of the graticule; and that the distances n which are tabulated as distances on the parallel, are actually laid off as abscissae at right angles to the meridian. For sheets not exceeding a degree square this divergence from the strict procedure is permissible; but it would not do for the sheets of the International Map; and it is not employed in the construction of the plane table graticule which follows.

TABLE VI. Specimen of Tables for construction of War Office Maps.

Coordinates of Projection for Degree Squares. Scale $\frac{1}{250000}$.

Latitude	m on Meridian Inches	n on Parallel Inches	q on Diagonal Inches	Ordinates of Curvature. Inches	
				At 15′	At 30′
52° 0′		10·816		0·005	0·019
	4·381				
15′		10·756			
	4·381				
30′		10·695			
	4·381				
45′		10·634			
	4·381				
53° 0′		10·574		0·005	0·018
	17·524		20·530		
53° 0′		10·574		0·005	0·018
	4·382				
15′		10·512			
	4·382				
30′		10·451			
	4·382				
45′		10·389			
	4·382				
54° 0′		10·328		0·005	0·018
	17·527		29·406		

Coordinates of Projection for Quarter Degree Squares. Scale $\frac{1}{125000}$.

Latitude	m on Meridian Inches	n on Parallel Inches	q on Diagonal Inches	Ordinate of Curvature. Inches At 15′
52° 0′		10·816		0·009
	8·762			
15′		10·756		
	8·762			
52° 30′		10·695		0·009
	17·524		20·561	
52° 30′		10·695		0·009
	8·762			
45′		10·634		
	8·763			
53° 0′		10·574		0·009
	17·525		20·499	

Table VII is a specimen of the Tables for the projection of Graticules of Maps, as given in the official Text-book of Topographical Surveying. They differ from the foregoing in that they are not adapted for plotting in the drawing office, by the method of rectangular coordinates, but are for use in the field, where the drawing facilities are limited to the plane table as a drawing board, and the sight rule for straight edge, with a pair of dividers.

The tables therefore give the sides and diagonals of the trapeziums which make up the graticule. The central meridian is drawn and divided into the lengths m. And the corners of the trapeziums are obtained by striking arcs from the points of division of this meridian, with radii equal to the sides n, and the diagonals q, as given in the tables. Finally, when the corners of the trapeziums are thus obtained, the sides are drawn as straight lines; and the parallels thus become series of straight lines, a polygonal representation of the arcs of the circles they should be, but a substitute amply good enough for the field work in question.

TABLE VII.

SPECIMEN OF TABLES FOR PLANE TABLE GRATICULES.

Graticules of Maps. Sides and Diagonals of Squares of a quarter of a degree of Lat. and Long. on the scale of 1 inch to 2 miles.

Latitude		Length in Inches			
		m	n	p	q
		on meridian	on lower parallel	on upper parallel	on diagonal
52° 0′ to 52° 15′		8·643	5·335	5·305	10·149
52 15	52 30	643	305	275	133
52 30	52 45	643	275	245	118
52 45	53 0	644	245	215	103
53 0	53 15	644	215	185	088
53 15	53 30	645	185	155	073
53 30	53 45	645	155	124	057
53 45	54 0	645	124	094	042

TABLE VII (*continued*).

Linear value in feet of one second of Arc and its Logarithm, measured along the Meridian.

Latitude	Length in feet	Logarithm	Diff.
52° 0′	101·4056	2·0060620	
			+63
5	4071	683	
			63
10	4085	746	
			62
15	4100	808	
			63
20	4115	871	
			62
25	4129	933	

Linear value in feet of one second of Arc and its Logarithm, measured along Parallels of Latitude.

Latitude	Length in feet	Diff.	Logarithm	Diff.
52° 0′	62·5932		1·7965271	
		−1163		−8076
5	4769		57195	
		1164		8100
10	3605		49095	
		1166		8125
15	2439		40970	
		1167		8149
20	1272		32821	
		1168		8174
25	0104		24647	
		1169		8200

Table VIII gives the radii in the principal zenithal projections for each 10° of true distance from the centre. A study of this table shows within what limits the different projections are fairly the same, and where they begin to differ widely one from another. If the figures here given be plotted on squared paper it is easy to derive intermediate values, and in this way it is simple to transform from one zenithal projection to another. For example, Table I gives the radii and azimuths for the zenithal equidistant projection. If we wish to construct the zenithal equal area we may take from such a diagram the radii of the latter which correspond to the already known radii of the former; and then, since the azimuths remain the same, it is a simple matter to pass from the one to the other.

TABLE VIII.

Comparison of the radii in different Zenithal Projections.

	Equi-distant	Equal area	Ortho-morphic	Breusing	Gnomonic	Ortho-graphic
10°	0·175	0·174	0·175	0·175	0·176	0·174
20	0·349	0·347	0·353	0·350	0·364	0·342
30	0·524	0·518	0·536	0·527	0·577	0·500
40	0·698	0·684	0·728	0·706	0·839	0·643
50	0·873	0·845	0·933	0·888	1·192	0·766
60	1·047	1·000	1·155	1·075	1·732	0·866
70	1·222	1·147	1·400	1·267	2·747	0·940
80	1·396	1·286	1·678	1·469	5·671	0·985
90	1·571	1·414	2·000	1·682	∞	1·000

The radius of the sphere is taken as unity.

Table IX gives the distances between the parallels on Mercator's projection, expressed in degrees of the equator. It will be noticed that the effect of the ellipticity of the Earth, in making a degree of longitude greater than a degree of latitude at the equator, at first more than balances the exaggeration in latitude which begins in this projection as soon as the equator is left.

TABLE IX.

Distance between the parallels on Mercator's Projection, taking into account the ellipticity of the Earth.

(The distances are expressed in degrees of the equator.)

0°—10°	9° 57′
10 —20	10 20
20 —30	10 59
30 —40	12 12
40 —50	14 9
50 —60	17 31
60 —70	23 57
70 —80	40 8
80 —85	39 50
85 —90	∞

Table X gives the distances from the equator of the parallels for every 10° of latitude, expressed in decimals of the semi-axis-minor of the ellipse. It will be remembered (see page 60) that this $= \sqrt{2} \times$ radius of the Earth on the desired scale. Thus if the total area of the map is to be that of a sphere whose radius is $1/10^8$ that of the Earth, $= 63·66$ mm., the quantities in the table must be multiplied by $63·66\sqrt{2}$ to give the values in millimetres.

TABLE X.

Mollweide's Projection. Distance of the parallels from the equator.

Lat.	Dist.
10°	0·137
20	·272
30	·404
40	·531
50	·651
60	·762
70	·862
80	·945
90	1·000

APPENDIX

By inadvertence no description of the Polyhedric projection was given in its proper place, though it is mentioned in the table on page 66 as in use for several series of topographical maps. The necessity of adding an appendix, to describe this projection, gives the opportunity of referring to two projections now in use by the Egyptian Survey Department. They are treated in Survey Department Paper No. 13, *The theory of Map Projections, with special reference to the projections used in the Survey Department*, by J. I. Craig, M.A., Cairo, 1910. The author is indebted to the Department for the gift of a copy of this paper.

Polyhedric projection.

This is of no scientific interest, except that it is much used in European maps. Take on the spheroid the four points that are to be the corners of the sheet, and pass a plane through them; this will be strictly possible if the sheet is to be bounded by meridians and parallels. Let fall perpendiculars from each point of the enclosed spheroidal trapezium to this plane, and we have the projection. The formulae for the calculation of coordinates are complicated, but in practice they are scarcely required, since within the limits of a single sheet not more than one degree square the projection is indistinguishable from the polyconic and from many other projections. Adjacent sheets fit along the edges, and the whole series of sheets representing a zone of latitude can be fitted together and laid out flat, but will not fit the adjacent zone. Thus it is not possible to combine a number of small sheets to make one large one, though in practice the difficulty would be felt only when it was a question of combining the original engraved plates. The deformations of the printed sheets would be much larger than those due to the projection alone.

The "Gauss Conformal projection."

This is the name given by Mr Craig (*loc. cit.*) to the projection which has been adopted for the whole of the maps of the Egyptian Survey. The name is confusing, since the Conical orthomorphic projection, which is not the same, is very frequently referred to as Gauss' projection; and the word "conformal" is the older equivalent of the modern word "orthomorphic."

This "conformal" projection was devised by Gauss for the survey of Hanover, but it was not properly described until Schreiber investigated its properties very fully in 1866. The peculiarity which distinguishes it from other orthomorphic projections is that the central meridian is true. The other meridians and the parallels are complex curves of which it is hard to give any geometrical account. The parallels diverge from one another on each side of the central meridian, and the scale in the north and south direction becomes wrong very quickly away from that meridian. Thus the projection is suitable only for a country such as Egypt, which is a narrow strip along a meridian. In this respect it resembles Cassini's projection, as used in the English Ordnance Survey, and it may perhaps be described in general terms as like a Cassini projection in which the meridians and parallels have been slightly modified so that they intersect everywhere at right angles and make an orthomorphic projection. [θ^*]

The expressions for the calculation of the rectangular coordinates are very complicated, and the whole theory quite unsuited for an elementary book such as this. A complete account of it is given by Mr Craig.

It is not easy to discover that the projection has any advantage over several others which are far simpler and of general applicability.

The Mecca Retro-azimuthal projection.

This is a very interesting example of a new class of projection, which may be described as the inverse of the ordinary azimuthal or zenithal projection. In the latter the azimuth of any point is true at the centre of the map. In the retro-azimuthal projection the azimuth of the centre is true at any point of the map. Thus if the map is centred on Mecca it is possible to find the true bearing of Mecca from any point. Such a map is of great interest to Mahometans in finding the direction of the " Qibla."

The general properties of the class have not yet been investigated, and it is not possible to give rules for constructing a retro-azimuthal equal area or orthomorphic projection. In the map produced by the Egyptian Survey Department the meridians are drawn as straight lines at their correct equatorial distances apart, and the intersections with the parallels are found by computing the azimuths of Mecca for the intersections, and laying them down so that these azimuths are true upon the map. A small reproduction of the map is given in a pamphlet published by the Egyptian Survey Department, Technical Lecture No. 3, 1908–09 : *Map Projections*, by J. I. Craig. [*κ**]

CHAPTER XI

NOTES IN CORRECTION AND AMPLIFICATION OF
THE PRECEDING CHAPTERS

A* (pp. v and 4). This is true of Atlas maps, but not of the larger scale maps of a topographical series. See Chap. XII, p. 136.

B* (p. 2). A fifth criterion should be applied to the projection of a series of topographical maps. Does it lend itself conveniently to the numerical work of the survey, the calculation of rectangular coordinates and of grids, and the publication of lists of coordinates? See Chap. XII, p. 133.

C* (p. 5). This again is true of Atlas maps, but not of topographical maps, though the advantage is not so much in the map itself as in the numerical records and calculations. On any single sheet of such a series the meridians will, however, be practically straight, though not accurately so. See Gauss Conformal Projection. Chap. XII.

D* (p. 7). This paragraph is beside the point. In practice the curved meridians and parallels are drawn through a series of plotted points, and it does not make very much difference whether they are circles or not.

E* (p. 8). In the compilation of maps, such as sheets of the 1/Million map, by selection of detail from many sheets on varying but larger scales, it is particularly advantageous that meridians and parallels should be at right angles throughout, and that all the material should be sensibly if not strictly orthomorphic. Reductions by photography or pantagraph to the scale of compilation will then be fitted with a minimum of trouble.

F* (p. 11). From the point of contact T of the tangent in Fig. 4 lay off any distance TA along the tangent and an equal arc *TB* on the circle. The length of the parallel of the projection through *A* is the circle whose radius is the perpendicular from *A* to the polar axis; and the length of the corresponding parallel of the sphere is the circle whose radius is the perpendicular from *B*. By drawing the figure one can see intuitively, though it is not so easy to prove it geometrically, that the former is always greater than the latter, and hence that the scale along any parallel other than the standard is greater than unity. See also the note to p. 79.

G* (p. 13). The theorem for the scale along the parallels may be considered thus: let r_1, r_2 be the radii of the standard parallels of latitude ϕ_1 and ϕ_2. For any intermediate parallel ϕ the radius on the projection is in simple proportion between r_1 and r_2. But owing to the convexity, the true length of an intermediate parallel on the earth is not in simple linear proportion to the lengths of the standard parallels, but is larger. Hence in the projection the scale along any parallel between the standard parallels is too small; while similar considerations show that outside it is too large.

H* (p. 17). The proofs of the theorems in this section which are "easy to show" will be found on pages 93–95.

K* (p. 20). As already noted in more than one place, this disparagement of the property of orthomorphism was overdone in the first edition of this book. See notes B*, C*, E*, and the treatment of projections for topographical maps in Chap. XII.

L* (p. 22). The "Gauss Conformal Projection," briefly described in the Appendix to the First Edition, p. 123, should have been mentioned here, as it is used in the Surveys of Hanover and Egypt. It is a transverse cylindrical projection : *i.e.* the cylinder may be imagined as touching the Earth along a meridian instead of along the equator (see Chap. XII). The oblique, not the transverse cylindrical projections are difficult to draw.

M* (p. 25). But we may note that if the simple cylindrical projection is made transverse it becomes Cassini's, which is of value, being used for the Ordnance Survey maps of England. See pp. 57 and 134.

N* (p. 37). These quantities, taken from the original paper by Airy (*Phil. Mag.* Dec. 1861, p. 409), were shown by Clarke to be seriously wrong, owing to an error in Airy's theory (*Phil. Mag.* Apr. 1862, p. 306). The corrected values are

	At 90° from centre	At 110°
Radius	20 % small	20 % large
Area	0·4 % small	420 % large
Scale along parallel	1·47 times	1·58 times

Scale along meridian.

O* (p. 38). When the projection is made on the tangent plane, the scale is correct in the immediate neighbourhood of the centre, but at the expense of the outlying parts of the map. By bringing the plane of projection in towards the centre we can distribute the errors better, and by a proper choice make the sum of the squares of the errors of scale, taken all over the area to be represented, a minimum for that area. See note on Clarke's projection, p. 48.

P* (p. 43). I am indebted to Mr R. M. Milne, of the Royal Naval College, Dartmouth, for the following direct construction:

Let P and Q be points on adjacent faces, which it is desired to join by a great circle. From Q draw QR perpendicular to the common edge, and produce it so that $RQ' = Q'T = QR$. (Fig. 20.)

Draw PT' perpendicular to the common edge, and join PQ', TT' intersecting in O. A line from L, the middle point of the edge, through O to cut the common edge in S gives the required point S such that PSQ is the great circle. For if SP, SQ' are produced to cut the edge through L in A, C', PQ' to cut the common edge in Z, and a line is drawn through O parallel to the edge to cut PT' in W, SP in K, $Q'T$ in V, and $Q'C$ in H, we have by similar triangles

$$\frac{OK}{SZ} = \frac{PW}{PT'} = \frac{Q'V}{Q'T} = \frac{Q'V}{Q'R} = \frac{OH}{SZ}.$$

Hence $OK = OH$ and $AL = C'L = CN$, which is the property required for PSQ a great circle.

Q* (p. 48). In the more general case, where the plane of

projection is not the tangent plane, but parallel to it at distance k from the centre of projection,

$$r = R \frac{k \sin \zeta}{h + \cos \zeta},$$

and the constants h and k have to be determined for the desired spherical radius β of the map. The scale is no longer true at

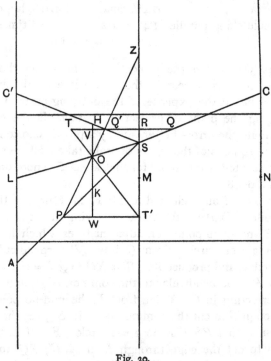

Fig. 20.

the centre, but the scale at the margins is considerably better, and the sum of the squares of the scale error taken over the map is much reduced. For a complete discussion, see the work on Map Projections by Mr A. E. Young published by the Royal Geographical Society as No. 1 of their new Technical Series.

R* (p. 52). The projection is usually so considered; but it is really the Transverse Simple Cylindrical. See note V*, p. 60, and Chap. XII.

S* (p. 53). Germain calls this the "Projection Sinusoïdale de Sanson improprement dite de Flamsteed," and his remark is just, for it was used by Sanson in 1650, for his first large terrestrial atlas, and by Flamsteed only after 1700 for his star-atlas published in 1727. It seems proper to call the projection by the name of Sanson alone. Monsieur d'Avezac still more justly resents an attempt made many years ago to call Bonne's projection the "modified Flamsteed."

T* (p. 56). This dismissal of the Polyconic projections as of no particular scientific interest is too summary. In Chap. XII it will be shown that for a map of an area such as Great Britain there is exceedingly little difference between the Polyconic, Cassini's, Lambert's Conical orthomorphic, and the "Gauss conformal" projections. In a new and unpublished investigation of the Rectangular Polyconic Mr G. T. McCaw, of the Geographical Section, General Staff, has shown how this projection can be modified with advantage to make it sensibly orthomorphic. (See forthcoming paper in *Geographical Journal.*)

U* (p. 58). Corrected tables have since been issued for insertion in the Report of the London Conference, 1909.

V* (p. 60). This judgment is also too summary: Cassini's projection, as remarked above, is the transverse simple cylindrical. In Chap. XII we shall see that by a slight modification it can be transformed into the "Gauss Conformal" or transverse Mercator projection. For a series of topographical maps it is of course necessary to take account of the Earth's ellipticity, which very much complicates the formulae. (See Chap. XII, p. 135.)

W* (p. 60). Unless they are to be provided with a rectangular grid for military use, in which case the sheets must remain rectangular, and the sheet lines will have a progressive deviation from the meridians and parallels. But bearings are then given from "grid north" instead of from true north.

X* (p. 62). It has been shown by Mr A. E. Young that Breusing's projection may be improved by taking the Harmonic mean instead of the Geometrical, and by altering the scale value so as to give total minimum error for a map of given radius, as in Airy's projection.

If β is the spherical radius of the map, we then have

$$r = 4 \frac{\sin^2 \tfrac{1}{4}\beta}{\log \sec \tfrac{1}{2}\beta - \tan^2 \tfrac{1}{4}\beta} \cdot \tan \tfrac{1}{4}\zeta.$$

Y* (p. 70). The conical projection with two standard parallels has been very much used for the maps published in the *Geographical Journal*, and the parallels made standard are always named in a note on the lower margin of the map.

Z* (p. 71). The parallels are curved but are not concentric circles.

a* (p. 72). But in choosing the projection for a series of topographical maps of a Survey Department we have to consider the convenience of the computing office as well as that of the drawing office, and also whether the sheets are to carry a rectangular grid for military use.

β* (p. 73). This now requires modification. An inter-Allied conference on Flying-maps has recently proposed that a series of scale 3 cm. to 1° Eq. shall be drawn on Mercator's projection for use in air-navigation. But there must be a better solution than to re-draw land maps on Mercator's projection for the exclusive use of airmen. They may learn to measure bearings from a map on which the meridians are convergent.

Further, a critic has pointed out that Mercator's projection is suitable for a land map of a strip along the equator, with little extent in latitude, just as the transverse Mercator is used in Egypt.

γ* (p. 76). But the transverse cylindrical, simple and orthomorphic, are important, and are dealt with fully in Chap. XII.

δ* (p. 79). In note F* (p. 11) we gave a geometrical illustration by which it is easy to see intuitively, though not to prove strictly, that the scale along any parallel other than the standard is too great. This may be shown analytically as follows:

ϕ and ϕ_0 are both numerically $< \pi/2$.

CASE I. If ϕ_0 is positive and ϕ is $> \phi_0$, $\sin \phi + \sin \phi_0 > 2 \sin \phi_0$.

$\therefore 2 \sin \tfrac{1}{2}(\phi + \phi_0) \cos \tfrac{1}{2}(\phi - \phi_0) > 2 \sin \phi_0,$

$2 \sin \tfrac{1}{2}(\phi + \phi_0) \sin \tfrac{1}{2}(\phi - \phi_0) > 2 \sin \phi_0 \tan \tfrac{1}{2}(\phi - \phi_0),$

$\cos \phi_0 - \cos \phi > \sin \phi_0 . 2 \tan \tfrac{1}{2}(\phi - \phi_0),$

and *a fortiori* $> \sin \phi_0 (\phi - \phi_0)$, since the tangent of an angle is always greater than the angle.

Hence $\cos \phi_0 - (\phi - \phi_0) \sin \phi_0 > \cos \phi$.

CASE II. If ϕ_0 is positive and ϕ is $< \phi_0$, we write

$\cos \phi_0 - (\phi - \phi_0) \sin \phi_0$

$= \cos \phi + \cos \phi_0 [1 - \cos (\phi - \phi_0)] + \sin \phi_0 [\sin \overline{\phi - \phi_0} - (\phi - \phi_0)].$

The two last terms are both positive. Hence

$\cos \phi_0 - (\phi - \phi_0) \sin \phi_0 > \cos \phi.$

If ϕ_0 is negative the cases are interchanged; but symmetry shows that the proposition must be true, and it need not be proved for this case separately.

The proof of Case I was suggested to me by Lt.-Col. Stratton, D.S.O., of Gonville and Caius College; and of Case II by Mr J. Jackson of Trinity College and the Royal Observatory, Greenwich.

ϵ^* (p. 96). This projection was much used during the war for the French large scale maps of the Western front, and during the autumn of 1918 it was being brought into use for all the Allied armies. For its calculation when the ellipticity of the Earth is taken into account, see Chap. XII.

ζ^* (p. 99). More properly, the name Equidistant implies only that the scale along the radial is uniform, not necessarily correct. In a valuable new work on Map Projections (Technical Series, No. 1, R.G.S.) Mr A. E. Young points out that if the radial scale be suitably diminished by an amount depending on the spherical radius of the map, the sum of the squares of the errors of scale in all directions, taken over the map, may be made a minimum, as in Airy's projection, and that when this is done the Zenithal Equidistant is practically as good as any zenithal projection that can be used. A similar modification may be made in the Zenithal Equal Area and Orthomorphic projections.

η^* (p. 103). But we had overlooked the transverse cylindrical orthomorphic or "Gauss conformal."

θ^* (p. 123). As noted above, it is better described as the Transverse Cylindrical Orthomorphic projection.

κ^* (p. 124). The retro-azimuthal projection as described by

Mr J. I. Craig has the meridians drawn as equidistant parallels, as in the Mercator projection, which facilitates the measurement of azimuth by a compass rose and parallel rulers. If the term be extended to include projections with curved meridians, in which the reverse azimuths are not so conveniently measurable, then the Zenithal Orthomorphic (Stereographic) is an orthomorphic retro-azimuthal projection, as was pointed out to me by my old pupil, the late Captain G. P. Blake, Royal Welch Fusiliers.

CHAPTER XII

PROJECTIONS FOR TOPOGRAPHICAL MAPS
ON LARGE SCALES

In our previous chapters we have been concerned mainly with projections for atlas maps and topographical maps on smaller scales, such as 1/Million. These are compiled from topographical material on much larger scales, and the compiler works with material already drawn in map-form. His problem is to choose a projection suitable for the representation of the large area which he proposes to cover, and if he needs to take account of the ellipticity of the earth, he can do so very simply from the tabulated lengths of a degree of the meridian and the parallel in different latitudes.

For the maker of large scale maps in an extensive national survey, the problem is very different. The material with which he has to deal is largely numerical—the results of his survey operations of triangulation and traverse. And for the orderly conduct of his calculations he has to maintain a standard of numerical accuracy much higher than corresponds to a visible quantity on any single sheet. If he did not he would in the course of his work accumulate errors which in the end *would* become visible on the drawing. This high degree of numerical accuracy is essential also in a true cadastral survey, where it is required to lay down the coordinates of at least one point in each property; and in the last few years it has been found indispensable to the proper coordination of survey and artillery in the operations of stationary warfare.

To attain this accuracy the surveyor has not only to take account in his formulae of the Figure of the Earth, but he has to decide which of the many determinations of figure he shall employ. And the choice is complicated by the fact that the actual dimensions of the Earth are not known to the required degree of numerical accuracy. The choice is to some extent arbitrary; and there has not been any measure of agreement between the surveys of the world, which figure to adopt. This has produced a quite unnecessary degree of confusion in the various geodetic tables computed and employed in different countries, and the greatest caution is necessary if several different organisations have to work to a common end, as was found in the survey operations on the Western Front.

Figure of the Earth adopted by the surveys of different countries.

The following are the principal determinations of the semi-diameters and the flattening of the Earth:

		Equatorial semidiameters	Polar semidiameters	Flattening
Carte de France		6376985 *m.*	6356323 *m.*	1/308·64
Everest	1830	6377276	6356075	1/300·80
Airy	1830	6376542	6356236	1/299·33
Bessel	1841	6377397	6356079	1/299·15
Clarke	1858	6378293	6356618	1/294·26
Clarke	1866	6378206	6356584	1/294·98
Clarke	1880	6378249	6356515	1/293·47
Hayford	1910	6378388	6356909	1/297·0

Since Everest, Airy, and Clarke worked in feet, and the relation between the foot and the metre has been the subject of several re-determinations, the conversion into metres is not always consistently given by different authorities.

Now the French used the *Carte de France* figure until recently, when for their new map on 1/50,000 they adopted Clarke 1880. The Ordnance Survey use Airy: the Survey of India use Everest. The United States Coast and Geodetic Survey use Clarke 1866; the Geographical Section, General Staff and South Africa use Clarke 1858; and Central Europe uses Bessel. Hence the greatest caution must be used in combining numerical results from different surveys.

Cassini's projection, as calculated by the Ordnance Survey.

A recent paper of the Ordnance Survey, printed for departmental use in 1919—*The Mathematical Basis of the Ordnance*

Maps of the United Kingdom, by Major A. J. Wolff, D.S.O., R.E.
—gives an account of the formulae employed, from which the following is abridged:

Let ϕ_0 be the latitude of the centre of the projection, and ϕ that of a point B whose coordinates are to be calculated, $\Delta\lambda$ being the difference of longitude between A and B. Draw the arc BM perpendicular to the central meridian on the spheroid. Then the spheroidal coordinates BM, AM, plotted on a plane as rectangular coordinates, are the coordinates of B in the Cassini projection with centre A.

Let ν be the radius of curvature at right angles to the meridian in latitude ϕ, and ρ the radius of curvature in the meridian.

The latitude of M is greater than ϕ by a small quantity η which we have to calculate.

To a first approximation

$$BM = x = \Delta\lambda . \nu \sin 1'' \cos \phi$$

and $$\eta = x^2 \tan \phi / 2\rho\nu \sin 1''.$$

More accurately,

$$x = \Delta\lambda . \nu_1 \sin 1'' \cos (\phi + \tfrac{1}{3}\eta),$$

and $$\eta = x^2 \tan (\phi + \eta)/2\rho\nu \sin 1'' - x^4 \frac{[1 + 3 \tan^2 (\phi + \eta)] \tan (\phi + \eta)}{24\nu^2\rho^2 \sin 1''},$$

where ν_1 is the radius of curvature at right angles to the meridian in latitude $\phi + \eta$.

This gives a second approximation to η.

Finally $$x = \Delta\lambda\nu_1 \sin 1'' \cos (\phi + \tfrac{1}{3}\eta),$$

$$y = (\phi + \eta - \phi_0) \rho_1 \sin 1'',$$

where ρ_1 is the radius of curvature of the meridian in latitude

$$\tfrac{1}{2}(\phi_0 + \phi + \eta).$$

It is sufficiently evident how much the calculation is complicated by taking into account the spheroidal shape of the Earth.

Alternatively, the rectangular coordinates may be calculated from the expansions in series given by Jordan, *Handbuch der Vermessungskunde*, vol. III, p. 457, cf. the Fifth Edition, 1907, for which see later: a great advantage of this method is that it is easy to see just where the particular properties of the projection become significant: or where it begins to diverge from other projections which might have been used for the same sheet, as Lambert's Conical Orthomorphic, the Polyconic, or the Transverse Cylindrical Orthomorphic.

Lambert's Conical Orthomorphic with two Standard parallels.

The theory of this projection (neglecting the ellipticity of the Earth) has already been given on p. 93. The projection had been used occasionally for Atlas Maps, and the well-known Russian 10-verst map (1/420,000) is reputed to be an example of it: but it had not been much used, though its interesting theory had been studied very thoroughly by Pizzetti (*Trattato di Geodesia Teoretica*, Bologna, 1905), until it suddenly became famous during the War. The French found themselves handicapped by the variable inclination of meridian to parallel in the projection of Bonne, on which nearly all pre-war maps of the *Service Géographique* were constructed, and they realised that an orthomorphic projection was required for ready cooperation between the survey sections and the artillery. The effective use of the artillery in night harassing fire and in surprise concentrations demanded that the ranges and bearings should be calculated from rectangular coordinates of guns and targets supplied by the survey sections. Such coordinates must for convenience be referred to one general system covering the entire front, to which system the rectangular "grid" overprinted for references on all maps of the battle area must also conform.

The earlier history of the adoption of Lambert's Conical Orthomorphic projection by the French army is not known to the writer; but in the early summer of 1918 he became interested in the various problems presented by the French lithographed table of rectangular coordinates and the kilometric grid (*carroyage*) of maps of the Western Front, which on the unification of the Allied Command were adopted by all the armies of the Allies. The French, working in the centesimal system, calculated the constants of the projection for one standard parallel 55^G, and then applied a constant scale reduction with the intention of making parallels 58^G and 52^G their true length, retaining the constant of the cone calculated for 55^G. This, though a convenient rapid approximation, and practically as good as the more strictly correct process given below, does not give precisely the same results as the rigorous theory when tables are calculated to eight figures; and there were certain other minor defects in the theory which made it difficult to reproduce the French results to the last significant figure. (But see page 149.)

Meanwhile the Americans, on their entry into the field, took up the study of the Lambert projection, and published, in the first instance, tables on the French formulae but with Clarke's figure of 1866 instead of the *Ellipsoïde de Plessis* used by the French: the results were naturally divergent.

A small pamphlet by the author: *Notes on the " Tables de Projection (Système Lambert) and on the kilometric grid (carroyage) for maps of the Western Front"* examines the somewhat minute and complicated questions involved in the French and American tables. It has not been published, but a certain number of copies are at the disposal of those interested. One may draw from it the conclusion that a thorough study of the map projection, and careful tabulation from strictly rigorous theory, is an important part of the preparation for modern war.

The rigorous theory of the Lambert Projection.

If r is the radius in the projection of any parallel whose co-latitude is χ, and e is the eccentricity of the spheroid; n the " constant of the cone "; and K a scale-value constant:

$$r = K (\tan \tfrac{1}{2}\chi)^n \cdot \left(\frac{1 + e \cos \chi}{1 - e \cos \chi}\right)^{\frac{ne}{2}} \quad \text{[Germain, p. 57]},$$

K and n are at present indeterminate.

If the co-latitudes of the chosen standard parallels are χ_1 and χ_2, n is determined from

$$n = \frac{\log \nu_2 \sin \chi_2 - \log \nu_1 \sin \chi_1}{\log \tan \tfrac{1}{2}\chi_2 \left(\frac{1 + e \cos \chi_2}{1 - e \cos \chi_2}\right)^{\frac{e}{2}} - \log \tan \tfrac{1}{2}\chi_1 \left(\frac{1 + e \cos \chi_1}{1 - e \cos \chi_1}\right)^{\frac{e}{2}}}$$

where ν_1 and ν_2 are the normals at χ_1 and χ_2 terminated by the minor axis or the radii of curvature at right angles to the meridian.

The quantity $\nu \sin \chi$ is frequently useful, and may very well be included among the tables it is convenient to calculate once for all. Alternatively we may write the numerator of n as

$$\log \sin \chi_2 - \log \sin \chi_1 - \log (1 - e^2 \cos^2 \chi_2)^{\frac{1}{2}} + \log (1 - e^2 \cos^2 \chi_1)^{\frac{1}{2}},$$

or as

$$\log \sin \chi_2 - \log \sin \chi_1 + \tfrac{1}{2} M e^2 (\cos^2 \chi_2 - \cos^2 \chi_1) + \tfrac{1}{4} M e^4 (\cos^4 \chi_2 - \cos^4 \chi_1)$$

where M is the modulus of the logarithms. The latter expression (suggested by Young) is convenient in calculation.

The denominator of n may be treated in several ways. If we write

$$\tan \tfrac{1}{2} (\chi + \Delta\chi) = \tan \tfrac{1}{2} \chi \left(\frac{1 + e \cos \chi}{1 - e \cos \chi} \right)^{\frac{e}{2}},$$

then $\Delta\chi$ is very nearly, but not quite, the same as the reduction from geographic to geocentric latitude. It is usual to suppose that it is good enough to assume the identity of the two quantities; but the difference made in doing so is not negligible. The geographical latitude reduced by $\Delta\chi$ is called the "isometric" latitude.

The reduction to geocentric latitude is

$$\tfrac{1}{2} e^2 \sin 2\chi + \tfrac{1}{2} e^4 \sin 2\chi \cos^2 \chi$$

(Craig, p. 10). Care must be taken that the second term is included in the calculation of any tables used for this reduction; it may amount to $3''$. By expanding the above expression for $\tan \tfrac{1}{2} (\chi + \Delta\chi)$ Young has shown that

$$\Delta\chi = \tfrac{1}{2} e^2 \sin 2\chi + \tfrac{5}{12} e^4 \sin 2\chi \cos^2 \chi,$$

when the second term is less by one-sixth part than the second term in the reduction to geocentric latitude. The maximum value of the difference is $0''\cdot 50$ for $\chi = 30°$.

The denominator may therefore be calculated from the original expression, which is not so troublesome as it looks, if an arithmometer is available for multiplying the logarithms by e;

or we may use the expansion for $\Delta\chi$; or we may, following Young, use the expansion for the complete denominator

$$\log \tan \tfrac{1}{2}\chi_2 - \log \tan \tfrac{1}{2}\chi_1 + Me^2 (\cos \chi_2 - \cos \chi_1) \\ + \tfrac{1}{3} Me^4 (\cos^3 \chi_2 - \cos^3 \chi_1).$$

In any case, as both numerator and denominator involve sums of differences, it is not possible to calculate n with certainty to the same number of significant figures as may be achieved in other parts of the calculation.

Finally, the scale constant K is obtained from

$$K = \frac{\nu_1 \sin \chi_1}{n (\tan \tfrac{1}{2}\chi_1)^n \left(\dfrac{1 + e \cos \chi_1}{1 - e \cos \chi_1}\right)^{\tfrac{1}{2}ne}} = \frac{\nu_2 \sin \chi_2}{n (\tan \tfrac{1}{2}\chi_2)^n \left(\dfrac{1 + e \cos \chi_2}{1 - e \cos \chi_2}\right)^{\tfrac{1}{2}ne}}$$

$$= \frac{\nu_1 \sin \chi_1}{n (\overline{\tan \tfrac{1}{2}\chi_1 + \Delta\chi_1})^n} = \frac{\nu_2 \sin \chi_2}{n (\overline{\tan \tfrac{1}{2}\chi_2 + \Delta\chi_2})^n}.$$

The scale on any parallel of radius r is $\dfrac{nr}{\nu \sin \chi}$.

Note that for the standard parallels χ_1 and χ_2 the radii are given very simply by

$$r_1 = \frac{\nu_1 \sin \chi_1}{n} \text{ and } r_2 = \frac{\nu_2 \sin \chi_2}{n},$$

whence it is possible to calculate quite easily the radii for these two parallels and verify the results of any approximations in series.

It is shown by Pizzetti (*Trattato di Geodesia Teoretica*, Bologna, 1905, p. 390) that if r_0 is the radius of the middle parallel, co-latitude χ_0 and r the radius of the parallel whose meridian distance from the middle parallel is m, reckoned positive for increasing distance from the pole of the projection, then for the projection on the tangent cone

$$r - r_0 = m + \frac{m^3}{6\rho_0\nu_0} - \frac{m^4 \cot \chi_0}{24\rho_0\nu_0^2} \left(1 - \frac{4e^2 \sin^2 \chi_0}{1 - e^2}\right) + \dots,$$

where ρ_0 and ν_0 are the radii of curvature along and at right angles to the meridian in co-latitude χ_0. The term in m^4 is not inappreciable.

When it is required to reduce to the projection with two

standard parallels, we must remember that, not only the scale value, but also the constant of the cone should be altered, and this requires the introduction into the expansion of another term of the same order as the last; it is important to note, however, that this additional term tends to cancel the above term in m^4. Consider for simplicity the sphere, of radius R, putting $e = 0$ in the above. On the tangent cone

$$r - r_0 = m + \frac{m^3}{6R^2} - \frac{m^4 \cot \chi_0}{24R^3}.$$

If two parallels equidistant d from the original are made standard it is shown by Young that

$$r - r_0 = m + \frac{m^3}{6R^2} - \frac{md^2}{2R^2} - \frac{m^4 \cot \chi_0}{24R^3} + \frac{m^2 d^2 \cot \chi_0}{12R^3} + \dots.$$

The term in md^2 is the equivalent of the scale correction employed sometimes to reduce from one to two standard parallels. The term in $m^2 d^2$ arises from the alteration in n required to make this process complete: an alteration which should not be neglected. For the spheroid the expansions are somewhat complicated, but it is evident that the effects will be closely the same. The approximate method gives a spacing of the parallels much nearer the rigorous solution than would seem likely at first sight.

Calculation from the Rigorous Theory.

(a) *Constant of the cone, n.*

Taking standard parallels of co-latitudes $\chi_1 = 43^g$, $\chi_2 = 47^g$, and the values of $\log a$ and $\log e^2$ from the French Tables, we have

$$\log \nu_1 \sin \chi_1 = 6\cdot60148989$$
$$\log \nu_2 \sin \chi_2 = 6\cdot63337698$$

and their difference, $=$ numerator of n, $= 0\cdot03188709$.

Or, using Young's expansion and $\log M = \bar{1}\cdot63778431$

$$\log \sin \chi_2 = \quad \bar{1}\cdot82802314$$
$$- \log \sin \chi_1 \quad -\bar{1}\cdot79604860$$
$$\tfrac{1}{2} M e^2 (\cos^2 \chi_2 - \cos^2 \chi_1) \quad -0\cdot00008712$$
$$\tfrac{1}{4} M e^4 (\cos^4 \chi_2 - \cos^4 \chi_1) \quad -0\cdot00000033$$

$$\overline{0\cdot03188709} \quad \text{as above.}$$

For the denominator, using the expansion for $\Delta\chi$

$$\Delta\chi = (\tfrac{1}{2}e^2 \sin 2\chi + \tfrac{5}{12}e^4 \sin 2\chi \cos^2 \chi)\frac{200}{\pi},$$

we find for

	$\chi_1 = 43^g$	$\chi_2 = 47^g$
first term	$0^g \cdot 2009711$	$0^g \cdot 2050166$
second term	$\cdot 0006599$	$\cdot 0006047$
$\chi + \Delta\chi$	$0 \cdot 2016310$	$0 \cdot 2156213$
$\log \tan \tfrac{1}{2}(\chi + \Delta\chi)$	$\bar{1} \cdot 54771921$	$\bar{1} \cdot 58964654$
		$\bar{1} \cdot 54771921$
	denominator	$0 \cdot 04192733$

Or, using the complete expression

$\dfrac{1 + e \cos \chi}{1 - e \cos \chi}$	$1 \cdot 13395365$	$1 \cdot 12650787$
$\tfrac{1}{2}e \log \dfrac{1 + e \cos \chi}{1 - e \cos \chi}$	$0 \cdot 00219564$	$0 \cdot 00208057$
$\log \tan \tfrac{1}{2}\chi$	$\bar{1} \cdot 54552360$	$\bar{1} \cdot 58756598$
	$\bar{1} \cdot 54771924$	$\bar{1} \cdot 58964655$

whose difference gives denominator $\qquad 0 \cdot 04192731$

Or again, using Young's expansion

$\log \tan \tfrac{1}{2}\chi_2$	$\bar{1} \cdot 58756598$
$- \log \tan \tfrac{1}{2}\chi_1$	$- \bar{1} \cdot 54552360$
$- Me^2 (\cos \chi_2 - \cos \chi_1)$	$- 0 \cdot 00011463$
$- \tfrac{1}{3} Me^4 (\cos^3 \chi_2 - \cos^3 \chi_1)$	$- 0 \cdot 00000043$
denominator	$0 \cdot 04192732$

The three results agree well. Taking the last as mean of the three, we have

$$n = 0 \cdot 76053251 \qquad \log \bar{1} \cdot 88111778$$

but observe that this must be uncertain by a unit in the seventh place.

Had we used geocentric instead of isometric latitude we should have increased the second term in the expansion for $\Delta\chi$ by one-fifth. We have then

$\tfrac{1}{2}$(geoc. co-latitude)	$21^g \cdot 6008815$	$23^g \cdot 6028712$
$\log \tan$	$\bar{1} \cdot 54772064$	$\bar{1} \cdot 58964776$
whose difference is the denominator		$0 \cdot 04192712$

The numerator being the same, we have with the geocentric latitude

$$n = 0.76053614 \ \log \bar{1}.88111985$$

differing from the above by several units in the sixth place, although the difference between isometric and geocentric latitude is so small.

(b) *Radii of the standard parallels.*

With n for the isometric latitude

log ν sin χ	6·60148989	6·63337698
colog n	0·11888222	0·11888222
	6·72037211	6·75225920
radii	5252573·16	5652742·45
distance between parallels	400169·29	

With n for the geocentric latitude we have similarly

radii	5252548·13	5652715·49
distance between parallels	400167·36	

As we are working with 8-figure logarithms the radii are necessarily uncertain by a unit in the first place of decimals, even when care is taken to allow for the influence of the point in the tables showing that the last figure has been raised. Hence the distance between the parallels, computed thus with the most scrupulous care from the original formulae, is uncertain to about 0·2 metres. This simple calculation of the radii of standard parallels is an easy check on the general accuracy of any expansion adopted.

As a check on these radii of the standard parallels compute the scale value:

log n	$\bar{1}$·88111778	$\bar{1}$·88111778
log r	6·72037211	6·75225920
colog ν sin χ	$\bar{7}$·39851011	$\bar{7}$·36662302
	0·00000000	0·00000000

showing that the selected parallels are precisely standard.

And for the scale constant K we have, when

$n = 0.76053251$, $e = 0.08043202$ and $\frac{1}{2}ne = 0.03058603$,

$$
\begin{array}{lcc}
\log \nu \sin \chi & 6.60148989 & 6.63337698 \\
\text{colog } n & 0.11888222 & 0.11888222 \\
\text{colog} (\tan \tfrac{1}{2}\chi)^n & 0.34564408 & 0.31366948 \\
\text{colog} \left(\dfrac{1 + e\cos\chi}{1 - e\cos\chi}\right)^{\frac{1}{2}ne} & \overline{1}.99833014 & \overline{1}.99841766 \\
\hline
 & 7.06434633 & 7.06434634
\end{array}
$$

which is in satisfactory agreement.

(c) *Radii of other parallels.*

For the calculation of the radii of the other parallels, from the general expression $r = K \tan^n \tfrac{1}{2}\chi \cdot \left(\dfrac{1 + e\cos\chi}{1 - e\cos\chi}\right)^{\frac{1}{2}ne}$ we take as examples the parallels of co-latitude 42ᵍ and 45ᵍ.

	$\chi\ 42^{\mathrm{g}}$	45^{g}
$\dfrac{1 + e\cos\chi}{1 - e\cos\chi}$	1.13573606	1.13029273
$\log K$	7.06434633	7.06434633
$n \log \tan \tfrac{1}{2}\chi$	$\overline{1}.64597514$	$\overline{1}.67063739$
$\tfrac{1}{2}ne \log \dfrac{1 + \cos\chi}{1 - \cos\chi}$	0.00169072	0.00162690
	6.71201219	6.73661062
r	5152431.08	5452687.69

When the constant of the cone and the radii of the principal parallels have been calculated thus, the rectangular coordinates of the intersections of meridians and parallels may be calculated and tabulated for say each 10′ in the usual way. Thence by interpolation, which must be conducted strictly, with account of second and third differences, the latitudes and longitudes of the corners of the rectangular sheets and of any other points of the grid may be calculated: the edges of the sheets graduated in latitude and longitude: and the principal points of the survey, if calculated in these coordinates, may be plotted on the sheets. This provides a ready method of transferring to the Lambert projection material originally drawn on a different projection: as for instance the maps of France and Belgium, originally on Bonne's projection.

Choice of projection for a series of topographical maps.

The ruling consideration must be, whether or no the map is to be a squared map, with a system of rectangular coordinates continuous from sheet to sheet over its whole extent. Such a map has great merits in war, and in the internal work of the Survey department: but it has the demerit that the rectangular sheets become more and more askew to the meridians as the map is extended east and west from the centre: the Russian 40-verst map of Asia is a well-known example on a small scale. The familiar discordance between true and magnetic north, leading to many mistakes in popular use, is aggravated by the variation of "grid north," which even on the maps of the relatively small area covered by the British Ordnance Survey, may amount to several degrees, and which becomes very unsightly, though less easily overlooked, on a series of great extent in longitude.

Considering the needs of national defence paramount, and the scientific order of the survey an important secondary requirement, it seems best to adopt a well-studied system of orthomorphic projection in rectangular coordinates as the foundation of the topographical mapping for any series, however extensive, but to bear in mind that popular needs will demand a second series of maps, transferred from the original plates to sheets bounded by meridians and parallels. This is relatively easy and inexpensive: the reverse process, to extemporise a squared map and system of rectangular coordinates in the sudden stress of war, is clearly undesirable.

. The choice of projection is then practically limited to that between Lambert's Conical with two standard parallels, for great extent of longitude, and the transverse Mercator, or so-called "Gauss Conformal" for great extent of latitude. Cassini's projection differs from the latter by a small term in the x-coordinates, readily applied. For the secondary sheets, bounded by meridians and parallels, there is practically nothing to choose between the polyconic and the others. Within the limits of a single sheet they are indistinguishable to the eye.

Expansion in series of the principal Projection formulae.

The expressions for the various projections are so different in their mathematical form that until they are expanded in series it is hard to see at what point they begin to differ numerically. It will therefore be useful to give the following expansions, taken originally from Jordan's work, with additions by Mr A. E. Young.

Let ϕ_0 be the latitude of the centre of the map; ρ and ν the radii of curvature in and perpendicular to the meridian at this point; $\eta^2 = e^2 \cos^2 \phi_0 / 1 - e^2$, where e is the excentricity of the figure of the Earth;

then the coordinates x and y, perpendicular and parallel respectively to the central meridian of a point distant $\Delta\lambda$ and $\Delta\phi$ in circular measure of longitude and latitude respectively from the centre, are as follows:

Cassini's Projection.

$$x = \quad \Delta\lambda \cdot \nu \cos\phi_0 \quad \dots \dots (1).$$
$$- \Delta\lambda\Delta\phi \cdot \rho \sin\phi_0 \quad \dots \dots (2).$$
$$- \Delta\lambda\Delta\phi^2 \cdot \tfrac{1}{2}\rho \cos\phi_0 (1 + \eta^2 + 3\eta^2 \tan^2\phi_0)/(1 + \eta^2) \dots (3).$$
$$- \Delta\lambda^3 \cdot \tfrac{1}{6}\nu \sin^2\phi_0 \cos\phi_0 \quad \dots \dots (4).$$
$$- \Delta\lambda^3\Delta\phi \cdot \tfrac{1}{6}\nu \sin\phi_0 \cos^2\phi_0 (2 - \tan^2\phi_0) \quad \dots \dots (5).$$
$$+ \Delta\lambda\Delta\phi^3 \cdot \tfrac{1}{6}\rho \sin\phi_0 \quad \dots \dots (6).$$
$$+ \dots \dots$$
$$y = \quad \Delta\phi \cdot \rho \quad \dots \dots (7).$$
$$+ \Delta\phi^2 \cdot \tfrac{3}{2}\nu \tan\phi_0 \cdot \eta^2/1 + \eta^2 \quad \dots \dots (8).$$
$$+ \Delta\lambda^2 \cdot \tfrac{1}{2}\nu \sin\phi_0 \cos\phi_0 \quad \dots \dots (9).$$
$$+ \Delta\lambda^2\Delta\phi \cdot \tfrac{1}{2}\rho \cos^2\phi_0 (1 + \eta^2 - \tan^2\phi_0) \quad \dots \dots (10).$$
$$+ \Delta\phi^3 \cdot \tfrac{1}{2}\rho (1 - \tan^2\phi_0) \cdot \eta^2/(1 + \eta^2)^2 \quad \dots \dots (11).$$
$$- \Delta\lambda^2\Delta\phi^2 \cdot \nu \sin\phi_0 \cos\phi_0 \quad \dots \dots (12).$$
$$+ \Delta\lambda^4 \cdot \tfrac{1}{24}\nu \sin\phi_0 \cos^3\phi_0 (5 - \tan^2\phi_0) \quad \dots \dots (13).$$
$$+ \dots \dots$$

Polyconic Projection.

x is the same to this order of small quantities;

y differs only in term (13) which is

$$- \Delta\lambda^4 \cdot \tfrac{1}{24}\nu \sin^3 \phi_0 \cos \phi_0.$$

Hence to transform from Cassini to the Polyconic we have only to apply to y the correction

$$- \tfrac{5}{24}\nu \sin \phi_0 \cos^3 \phi_0 \cdot \Delta\lambda^4.$$

Lambert's Conical Orthomorphic with one standard parallel, that of the centre.

The x-coordinate differs in terms (3) and (5), which are respectively $- \Delta\lambda\Delta\phi^2 \cdot \tfrac{3}{2}\rho \cos \phi_0 \tan^2 \phi_0 \eta^2/(1 + \eta^2)$

and $\qquad - \Delta\lambda^3\Delta\phi \cdot \tfrac{1}{6}\nu \sin \phi_0 (1 + 4\eta^2 - 3\eta^2 \tan^2 \phi_0)/(1 + \eta^2)^2$

and the y-coordinate in terms (10), (11) and (13), which are respectively $- \Delta\lambda^2\Delta\phi \cdot \tfrac{1}{2}\rho \sin^2 \phi_0,$

$$+ \Delta\phi^3 \cdot \tfrac{1}{6}\rho (1 + 4\eta^2 - 3\eta^2 \tan^2 \phi_0)/(1 + \eta^2)^2,$$

and $\qquad - \Delta\lambda^4 \cdot \tfrac{1}{24}\nu \sin^3 \phi_0 \cos \phi_0.$

Hence to transform from Cassini to Lambert we add to term (3) the correction $+ \Delta\lambda\Delta\phi^2 \cdot \tfrac{1}{2}\rho \cos \phi_0$; substitute the new term for (5); add to (10) the correction

$$- \Delta\lambda^2 \cdot \Delta\phi \cdot \tfrac{1}{2}\rho \cos^2 \phi_0 (1 + \eta^2);$$

add to (11) the correction $+ \Delta\phi^3 \cdot \tfrac{1}{6}\rho/1 + \eta^2$; and to (13) the correction $- \Delta\lambda^4 \cdot \tfrac{5}{24}\nu \sin \phi_0 \cos^3 \phi_0$; all of which are small and readily calculated.

The projection thus derived from Cassini's may be readily, by a change of scale, be made to have two standard parallels; but if it be desired to make two strictly chosen parallels standard, we must proceed from the original formulae.

Gauss Conformal.

The x-coordinates differ in term (4) which is

$$- \Delta\lambda^3 \cdot \tfrac{1}{6}\nu \cos^3 \phi_0 (\tan^2 \phi_0 - 1 - \eta^2),$$

and in term (5) where the last factor is $5 - \tan^2 \phi_0$. To transform from Cassini we have therefore to add as corrections

$$+ \Delta\lambda^3 \cdot \tfrac{1}{6}\nu \cos^3 \phi_0 (1 + \eta^2) - \Delta\lambda^3\Delta\phi \cdot \tfrac{1}{2}\nu \sin \phi_0 \cos^2 \phi_0.$$

The y-coordinates are the same in both projections.

CHAPTER XIII

NOTES ON THE HISTORY OF MAP PROJECTIONS

THE invention of the Gnomonic projection is ascribed to Thales (B.C. c. 600) and of the Orthographic and Stereographic to Hipparchus (B.C. c. 150).

Ptolemy in his *Geography* (A.D. c. 150) describes two projections for the World Map, which are approximations to the Simple Conic and to the projection now generally known as Bonne's. In the various manuscripts and early printed editions of his work the World Map is impartially on either of these. The conical projection has the peculiarity that the limiting southern parallel (about 16° S.) is made the same length as the corresponding parallel north, and the meridians east and west of the central are bent suddenly at the equator to conform. As knowledge of the world increased at the end of the fifteenth century, and it became necessary to extend the World Map, this defect became intolerable, and the first type fell into disuse, while the second was developed by Waldseemüller in his celebrated map of 1507 into the heart-shaped type, of which the World Map of Apianus (1520) is the most perfect example. All maps of this class are extreme examples of Bonne's, or its particular case with parallels centred on the pole described by Werner (1514).

The extreme obliquity of the meridians to the parallels at the eastern and western limits led naturally to the division into two hemispheres, of which the first, though incomplete example, is found on the inset to Waldseemüller's map.

In the 26 other maps of Ptolemy's Atlas the projection is also of two types: either the cylindrical with standard parallel approximating to that of Rhodes, or in a type usually ascribed to Dominus Germanus (Donis), a polygonal projection made by

10—2

drawing the top and bottom parallel as straight lines, dividing them truly, and joining the points of division to make the meridians: a projection that was often employed in the atlases of the 16th and 17th centuries but has nothing to recommend it.

The first serious improvement in an atlas projection is due to Mercator, who for his large map of Europe (1554) employed the conical projection with two standard parallels usually ascribed to Delisle, but singularly neglected until recent years. The more famous projection of Mercator was devised for his World Map of 1569, the construction and properties being briefly described in legends on the face of the map. Towards the end of the century the Cambridge mathematician Edward Wright examined the theory more fully and made improved tables, which he showed to Hondius in manuscript, but strongly resented the use which Hondius made of them in publishing a World Map. Hence apparently the care with which Hondius, in the legend on his great map of 1608, recently brought to light, gives an explanation of the construction quite different from Wright's (see *Geog. Journal*, vol. LIV, p. 181). The logarithmic formula which completely describes the projection was first given by Bond in 1645.

In the late 16th and early 17th centuries there were many conventional projections for the world map and the hemispheres, of which the interest is only historical. But at this time much use was made of the Stereographic projection for the hemispheres, as in the great map of Blaeu recently presented to the R. G. S. (*Geog. Journal*, vol. LV, p. 312), engraved about 1650, since when this projection has gradually fallen almost into disuse, being superseded by the Zenithal equidistant and equal area: the former used from the 16th century in its polar form; the latter due to Lambert, who in 1772 published that treatise on projections which was the most substantial contribution to the subject ever made by one man.

Meanwhile for Atlas maps the projection now generally known by the name of Bonne, though descended from Ptolemy and much employed from the time of Bernard Sylva (1511), came into general use, with its special case of straight parallels properly known by the name of Sanson (1650), though often called by the name of Flamsteed, who used it for his star charts half a century later ; these, with the simple conic also descended from Ptolemy, have served for the greater part of atlas maps until the present time. The more refined types of conical projection, such as Albers' equal area, or Lambert's orthomorphic, have found as little favour with atlas makers as have the zenithal projections of Airy or Clarke.

The first considerable topographical map was that of France, which occupied the Cassinis, third and fourth of that name, during the greater part of the 18th century. The projection they employed was superseded in France by the Projection du Dépôt de la Guerre (Bonne's) but was about the same time, early in the 19th century, adopted by the Ordnance Survey for England, though curiously enough the maps of Scotland and Ireland are on Bonne's projection, which has, in spite of its obvious defects, been as much employed for topographical as for atlas maps. In the middle of the 19th century, while the United States developed the Polyconic, and the British War Office its modification the Rectangular Polyconic, the states of Europe were mostly content with Bonne or with the uninteresting Polyhedric : only the surveys of Mecklenburg and of Hanover struck out new lines with the employment of Lambert's Conical orthomorphic and transverse cylindrical orthomorphic (Gauss conformal): the latter is now employed by the Survey of Egypt, and the former became suddenly famous during the Great War.

But there is a curious point to be noted here. Since the pamphlet mentioned on p. 137 was written, the author has received a letter from Colonel Bellot, Director of the Service Géographique de l'Armée, which throws much light on the history that was unknown when the paragraphs on pp. 136 and 137 were written. The French did not in fact intend to adopt precisely Lambert's Conical Orthomorphic Projection, but one closely resembling it: Tissot's projection of minimum deformation for the case of a zone, which is the same to terms of the third order. Nevertheless the tables were issued as *Tables de Projection* (*Système Lambert*) and a good deal of mystification has been caused thereby. The work of Monsieur A. Tissot, *Mémoire sur la représentation des surfaces* (Paris, Gauthier-Villars, 1881) has not received in England the attention which it deserves, and the neglect must be remedied in a future edition of this book.

Such in brief is the history of our subject. If it should appear that the conservatism of map makers has made them rather indifferent to the performances of the mathematicians, we may reflect that the latter have not always been good propagandists, perhaps because they have not generally had at command the resources in cartography to demonstrate the merits of their inventions. Too many interesting projections, as Clarke's, can be seen only in a solitary specimen of small size. That there is still room for invention and resource is shown by the admirable new Projection of the International Map, and by the Transverse Mollweide of Sir Charles Close which forms the frontispiece of this book. The ingenious student will find that there is still ample room for experiment in the adaptation of known projections to the small world map so often required in text books: but to get the continents well disposed he must be ready to tackle oblique projections.

CHAPTER XIV

ADDITIONAL TABLES

THE need which arose during the Great War for general maps of Africa and of Asia, in many sheets, led to the examination of the question, how far the principle of the Projection of the International Map could be employed for sheets of like dimensions, but on smaller scales. The result of the enquiry was to show that it was not good for a scale less than 1/Two million, the departure from rectangularity at the corners being too great for the satisfactory mounting of combined sheets. For a map of Asia, on the scale 1/Five million, it was better to use Lambert's Conical Orthomorphic, in spite of the inevitable variation of scale over so large an area.

The tables for the projection of these maps were calculated at the Royal Geographical Society by Mr A. E. Young and the author, and may be reproduced here for more general use.

TABLE XI (A, B, AND C).

TABLES FOR THE PROJECTION OF MAPS ON THE SCALE OF 1/TWO MILLION, BY AN EXTENSION OF THE PROJECTION FOR THE INTERNATIONAL MAP OF THE WORLD.

TABLE XI (A).

Natural and Corrected Lengths on the Central Meridian.

Latitude		Natural length of 8° on scale 1/2M : inches	Correction*	Corrected lengths for plotting : inches
From − 4° to + 4°		17·413	− ·048	17·365
0	± 8	·414	48	·366
± 4	±12	·416	47	·369
8	16	·420	46	·374
12	20	·426	44	·382
16	24	·433	42	·391
20	28	·442	40	·402
24	32	·451	37	·414
28	36	·462	34	·428
32	40	·474	31	·443
36	44	·486	28	·458
40	48	·498	25	·473
44	52	·510	21	·489
48	56	·523	18	·505
52	60	·535	15	·520
56	64	17·545	12	17·533

* This correction to the central meridian is − ·048 cos² ϕ, where ϕ is the mean latitude of the sheet. The meridians 4° on each side of the centre are thus made of their true length, and the error of the bounding meridians 6° from the centre is numerically equal, but of opposite sign, to that of the central meridian.

TABLE XI (B).

Coordinates in inches of the Intersections of the Meridians with the Top and Bottom Parallels.

Latitude		Longitude from the Central Meridian					
		1°	2°	3°	4°	5°	6°
0°	x	2·192	4·383	6·574	8·766	10·957	13·149
	y	·000	·000	·000	·000	·000	·000
4	x	2·186	4·372	6·558	8·744	10·930	13·116
	y	·001	·005	·012	·021	·033	·048
8	x	2·170	4·340	6·510	8·680	10·850	13·020
	y	·002	·010	·023	·042	·066	·095
12	x	2·144	4·288	6·432	8·575	10·718	12·862
	y	·004	·015	·035	·062	·097	·140
16	x	2·107	4·214	6·321	8·427	10·534	12·640
	y	·005	·020	·045	·081	·126	·182
20	x	2·060	4·120	6·180	8·239	10·298	12·357
	y	·006	·024	·055	·098	·153	·221
24	x	2·003	4·006	6·008	8·011	10·013	12·015
	y	·007	·028	·064	·114	·178	·256
28	x	1·936	3·872	5·808	7·744	9·679	11·613
	y	·008	·031	·071	·127	·198	·285
32	x	1·860	3·720	5·580	7·439	9·297	11·155
	y	·008	·034	·077	·137	·215	·309
36	x	1·775	3·550	5·324	7·098	8·871	10·643
	y	·009	·036	·082	·145	·227	·327
40	x	1·681	3·362	5·042	6·721	8·401	10·079
	y	·009	·037	·085	·151	·235	·339
44	x	1·579	3·158	4·736	6·313	7·890	9·465
	y	·009	·038	·086	·153	·239	·344
48	x	1·469	2·938	4·406	5·873	7·340	8·805
	y	·009	·038	·085	·152	·238	·343
52	x	1·352	2·704	4·055	5·405	6·754	8·103
	y	·009	·037	·083	·149	·232	·335
56	x	1·228	2·456	3·684	4·910	6·136	7·360
	y	·008	·035	·080	·142	·222	·319
60	x	1·098	2·196	3·294	4·391	5·487	6·582
	y	·008	·033	·074	·133	·207	·299

To construct the projection, plot the top and bottom parallels only from the coordinates of Table B, with distance apart on the central meridian from the last column of Table A. Join corresponding points by straight lines for the meridians, and divide the meridians equally to obtain points on the intermediate parallels. The table is computed for each four degrees of latitude, to provide for series of sheets bounded by $\pm 4°$, $\pm 12°$, $\pm 20°$, ... (the series under construction); or alternatively $0°$, $\pm 8°$, $\pm 16°$, $\pm 24°$, ..., which might be more desirable in other continents. Note that the middle parallel is not plotted from the tables, but constructed as above.

For sheets $12°$ by $8°$ the angle between an extreme and the central meridian is given in the following Table C; the difference θ between these angles on successive sheets measures the bend of the extreme meridian at the junction of two sheets fitted along a parallel; 2θ is the angle shown in the figure between the inner margins of two sheets of the top row; and $2 \times 17.5 \sin \theta$, approximately 0.01θ (where θ is in minutes of arc), is the separation in inches at the top inner corner. Finally, the last columns of the table show the shortening of the middle parallel due to making the extreme meridians straight, and the consequent error of scale along this central parallel.

TABLE XI (C).

Lat.		Inclin. of extreme meridian	Diff. $= \theta$	Separation	Shortening of central parallel	Scale error
4° to 12°		51''·2			0·062 in.	·0024
			47'·3	0·48 in.		
12	20	98''·5			·061	·0024
			48''·9	0·50		
20	28	147''·4			·060	·0025
			43'·7	0·45		
28	36	191''·1			·054	·0025
			41''·4	0·42		
36	44	232'·5			·050	·0025
			34'·8	0·35		
44	52	267''·3			·042	·0024
			31''·7	0·32		
52	60	299'·0			·035	·0024

The figure shows the assemblage of four sheets plotted from these tables. Sheet I between meridians 18° and 30° E. and parallels 12° and 20° N. has length on central meridian 17·382 inches from Table A. The coords. of A with respect to M are $x = 12·357$, $y = +0·221$ inch; and of A' with respect to M' are $x = 12·862$, $y = +0·140$. Other points on the top and bottom parallels, at each degree of longitude, are plotted similarly from Table B. Corresponding points are joined by straight lines, and subdivided equally to obtain the other parallels. The meridian BB' 8° from the central meridian is its true length. Meridians AA' and MM' are respectively long and short by 0·044 inch (Table A). The meridian AA' is inclined 98'·5 to MM'; it makes an angle 48'·9 with the prolongation of the same meridian on the sheet above. The extremity C of the central parallel OC is 0·061 inch nearer O than it would have been if each parallel had been plotted independently on the polyconic projection (and the meridians therefore curved). The consequent error of scale on the middle parallel is 0·0024, the extreme parallels being correct. Finally, the separation at the top inner corner of sheets of the top row (near III) is 0·50 inch. All the latter figures are from Table C.

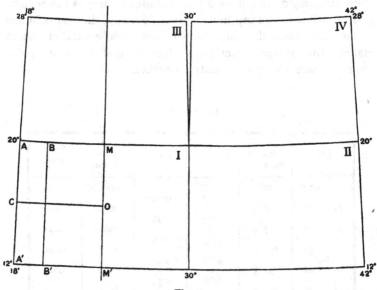

Fig. 21.

TABLE XII (A, B, C, and D).

TABLES FOR THE PROJECTION OF A GENERAL MAP OF ASIA, ON LAMBERT'S CONICAL ORTHOMORPHIC PROJECTION WITH STANDARD PARALLELS 63° N. AND 27° N.

Scale 1/5 Million.

Constants: log *a* 6·80470451   log *e*² 3̄·82754179
corresponding to flattening 1/297·0   log *n* 1̄·8571532.

TABLE XII (A).

Radii of Bounding Parallels for sheets covering 20° of latitude.

Lat.	Radius	Diff.
0°	2243·24 mm.	
		504·62
20	1738·62	
		439·15
40	1299·47	
		426·37
60	873·10	
		482·74
80	390·36	

TABLE XII (B).

Coordinates for Plotting Parallels: Millimetres.

Lat.		0°	20°	40°	60°	80°
ΔL 1°	*x*	28·18	21·84	16·32	10·97	4·90
	y	0·18	0·14	0·10	0·07	0·03
3°	*x*	84·51	65·50	48·96	32·89	14·71
	y	1·59	1·23	0·92	0·62	0·28
5°	*x*	140·80	109·12	81·56	54·80	24·50
	y	4·42	3·43	2·56	1·72	0·77
7°	*x*	196·99	152·68	114·11	76·67	34·28
	y	8·67	6·72	5·02	3·37	1·51
9°	*x*	253·06	196·13	146·60	98·50	44·04
	y	14·32	11·10	8·29	5·57	2·49
11°	*x*	308·97	239·47	178·98	120·26	53·77
	y	21·38	16·57	12·39	8·32	3·72
13°	*x*	364·69	282·65	211·26	141·94	63·46
	y	29·84	23·13	17·29	11·62	5·19
15°	*x*	420·17	325·65	243·40	163·54	73·12
	y	39·70	30·77	23·00	15·45	6·91

TABLE XII (C).

Distances of Parallels of even degree from the lower parallel: Millimetres.

Deg. from lower parallel	Lower Parallel			
	0°	20°	40°	60°
2°	55·30	45·90	42·39	44·04
4	109·29	91·22	84·69	88·63
6	162·09	136·01	126·96	133·87
8	213·77	180·34	169·24	179·89
10	264·41	224·24	211·59	226·82
12	314·09	267·77	254·06	274·82
14	362·89	310·98	296·71	324·11
16	410·87	353·91	339·60	374·94
18	458·09	396·61	382·80	427·66
20	504·62	439·13	426·38	482·74

TABLE XII (D).

Scale value of the Projection in different latitudes.

Lat.	Scale	Lat.	Scale
0°	1·266	20°	1·044
2	1·235	22	1·030
4	1·207	24	1·017
6	1·181	26	1·005
8	1·156	28	0·995
10	1·134	30	0·986
12	1·113	32	0·978
14	1·093	34	0·971
16	1·075	36	0·965
18	1·059	38	0·959
20	1·044	40	0·956
40°	0·956	60°	0·983
42	0·953	62	0·994
44	0·952	64	1·007
46	0·951	66	1·023
48	0·952	68	1·041
50	0·953	70	1·063
52	0·956	72	1·089
54	0·960	74	1·121
56	0·966	76	1·158
58	0·974	78	1·205
60	0·983	80	1·265

Note that between 20° and 70° the maximum scale error is about 5 per cent., and that the standard parallels were chosen for the best representation of the area between these parallels, which include the major part of the Continent.

INDEX

Printed in the United States
By Bookmasters